ASTRONAUT

**1961 onwards
(all roles and nationalities)**

COVER IMAGE Astronaut wearing the Extravehicular Mobility Unit (EMU), the American spacesuit used on the International Space Station. He wears the Primary Life Support Subsystem (PLSS) as a backpack. Attached to the sides and base of the PLSS is the Simplified Aid for EVA Rescue (SAFER) unit, which can propel the astronaut back to safety should they drift off. *(Ian Moores)*

First published in February 2017

A catalogue record for this book is available from the British Library.

ISBN 978 1 78521 061 7

Library of Congress control no. 2016958407

Published by Haynes Publishing,
Sparkford, Yeovil,
Somerset BA22 7JJ, UK.
Tel: 01963 440635
Int. tel: +44 1963 440635
Website: www.haynes.com

Haynes North America Inc.,
859 Lawrence Drive, Newbury Park,
California 91320, USA.

Printed in Malaysia.

Acknowledgements

I would like to thank Steve Rendle of Haynes for his unflagging enthusiasm, and my meticulous and indefatigable editor David M. Harland. The expertise of picture editor David Woods improved many illustrations and created several excellent graphics. James Robertson's layouts have skilfully balanced the text and illustrations.

I have had informative conversations and spent much entertaining time with many astronauts and cosmonauts over the years. Those with whom I have had the most helpful dialogue include Rusty Schweickart, Dorin Prunariu, Georgi M. Grechko, Edgar Mitchell, Leroy Chiao, Alan Bean, Vance Brand, Dick Gordon, Dave Scott, Bruce McCandless, Walt Cunningham, Al Worden, Anatoly P. Artsebarsky, Aleksandr Pavlovich Aleksandrov and the family of Viktor I. Patsayev.

Help with providing illustrations was generously given by Joachim Becker of Spacefacts.de, Richard Kruse of HistoricSpacecraft.com, Bob Gay of Famous and Forgotten Fiction, Colin Burgess, David Shayler, Tony Quine, Alan Bean, Lucy West, Ron Garan, Douglas Cooper, Chris Riley, Seán Doran, Ulli Lotzmann, Carl Walker, John Becklake and the British Interplanetary Society in London. Useful discussion on the Soviet and Russian space programs was had with the Memorial Museum of Cosmonautics in Moscow, Reginaldo Miranda Junior, Anatoly Zak of RussianSpaceWeb.com, and Phil Clarke. As always, the team from the Apollo Lunar Surface Journal and the Apollo Flight Journal were a regular source of unmatchable expertise. Best efforts have been made to track down the owners of images found on the internet.

Finally, I must acknowledge the support of my family, Caroline, Laura and Catriona, whose encouragement is occasionally mixed with affectionate scepticism about my various enthusiasms.

Terminology and units

Although there are influential exceptions among the scientific and engineering communities, the majority of English speakers use imperial units, so these are given first, followed by their metric equivalents (in parentheses). Many American space missions used the nautical mile for space distances, which is 1.15 statute miles (1.85 kilometres). Since most general readers do not use the nautical mile, this convention has not been followed.

This book is titled *Astronaut*, but the first man in space was a Russian cosmonaut. The two terms are used interchangeably, but cosmonaut is primarily employed when referring to space travellers in Soviet or Russian craft, irrespective of their nationality. Equivalent terms arising in other languages are not generally used in English, even if they have been anglicised.

The first astronauts were men, but women have now performed all astronaut roles, including pilot and commander. When used generally to refer to an astronaut or the role, 'he' also means 'she'.

ASTRONAUT

Haynes ®

1961 onwards
(all roles and nationalities)

Owners' Workshop Manual

An insight into the selection, training, equipment, roles and experiences of astronauts

Ken MacTaggart

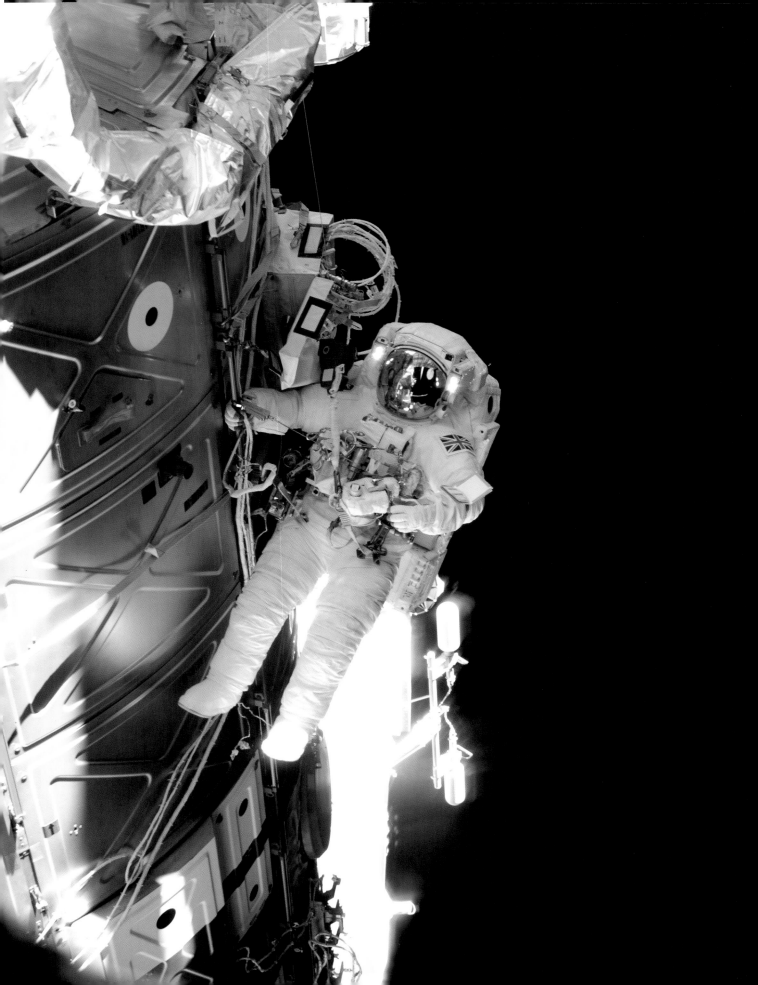

Contents

OPPOSITE **British astronaut Tim Peake spacewalking outside the International Space Station, January 2016.**

Foreword

by Leroy Chiao

Spaceflight was once mainly about heroic test pilots flying daring missions into the cosmos. Today, scientists and engineers make up the bulk of the different national contingents of astronauts, and those professions undertake most of the work on the International Space Station.

My own route into space began one hot summer afternoon on the shaded patio of our family home in California. With friends and family, I watched in awe on our fuzzy black-and-white television set as two white-suited figures took mankind's first steps on the Moon.

Many years of hard study and training later, I was privileged to be able to fulfil my boyhood dream of travelling to orbit, where I lived for a total of 229 days. I made four spaceflights with my American, Russian, Japanese and Italian crewmates over a 15-year career as a professional astronaut.

Human spaceflight is increasingly an international activity. As well as the three nations which have a demonstrated capability to send astronauts into orbit – the US, Russia and China – we now have Japan, Canada and the nations of the European Space Agency coming together on the Space Station.

With English as my first language, Mandarin Chinese as my second family tongue, and having lived in Russia to train for space, I see languages as being vital to a career in modern astronautics. It is a field in which international co-operation is increasingly important, and collaboration will be essential for the more ambitious astronaut missions now being planned.

Back on Earth, I have spent time with astronauts of many nations exchanging our impressions and responses to spaceflight. All of us have been transformed in some way by the experience of space travel.

I believe the perspective on the Earth afforded by the vantage point of space is the key to the future of the human species. No-one could fail to be overwhelmed by the fragile beauty of our planet and the tenuous hold on life it gives us, in the endless inhospitable blackness of space.

In time, some of the human race, perhaps most, will leave Earth for good. One day, perhaps we will all be astronauts.

__Dr Leroy Chiao__ is an American Space Shuttle astronaut who has also flown as a qualified co-pilot on Russian spacecraft and commanded the International Space Station. He was the first American to visit the Astronaut Research and Training Centre of China, where he compared experiences with China's first astronauts. He trained as a chemical engineer, worked on advanced materials and conducted medical experiments in orbit. He now holds appointments in academia and operates a training and education company.

Introduction

Once he was the personification of the ultimate pioneering explorer, heroically braving unknown dangers in mysterious corners of the universe. The fictional astronaut travelled the solar system and the galaxy to battle with evil opponents, saving Earth and the human race from some alarming fate. The space-suited figure was dashing, strong, courageous, skilful – and invariably male.

The first real astronauts of Russia and the USA were recruited from their respective armed forces. In the nervous tension of the Cold War, with its threat of nuclear annihilation, they became the public face of a more peaceful competition in space, which itself was something of a surrogate for the global struggle of political ideas then underway.

Those early cosmonauts and astronauts were filtered out of the military for mysterious 'special missions'. This was no planned career path, but a secretive, somewhat random process, whereby the hand of the state descended and plucked the perplexed military officer away for exclusive training. At the beginning, they were often not allowed to tell even their families what they were working on.

After their spaceflights, and sometimes even before they accomplished anything, the early American astronauts and Russian cosmonauts were fêted by the public, treated as national heroes and adulated as celebrities. Many were reluctant figureheads and recoiled from the overbearing attention but a few, whether they wished it or not, have acquired truly historic stature. The names of the first man in space and the first man to walk on the Moon are sure to be remembered a thousand years from now.

Now, however, the first phase of pioneering explorations of space is over. The once-exotic and elite designation of astronaut is becoming the job title of a career. It may not yet be routine work, but it is a form of employment that is becoming ever more accessible. Today, many men and women have deliberately set out to become astronauts as their chosen career path, and succeeded.

Posts are advertised, individuals can apply, candidates go through a comprehensive selection process and if successful they get the job, like any terrestrial employment. The final step of actually flying in space requires a high standard of performance in a wide range of further training, but the opportunity is in theory open to any person who fits the admittedly rigorous qualifying criteria.

This strict process, and the limited number of nations currently having the means to access space, restricts openings. Nevertheless, around 600 individuals from a wide range of backgrounds have already travelled in space.

All that is about to change, and an explosion of astronaut opportunities will emerge in the coming years as private sector spaceflight and space tourism are expected to expand dramatically. The era of elite astronauts being employed and trained exclusively by governments, by either the military or civilian space agencies, came to an end in 2004. The privately funded, sub-orbital flight of SpaceShipOne marked the creation of a new type of space traveller, the commercial astronaut.

Space is now a permanent habitation zone for the human species. Except for a few months in 2000, there has been a continuous human presence in space since 1990, shared between the Russian Mir space station and today's International Space Station.

The space sector of the global economy, which has the astronaut as its most exotic job specification but also a myriad of other roles, is already worth several hundred billion dollars annually. Many companies have plans to open space up to a wider public, and a few non-professional space travellers, known officially as *spaceflight participants*, have already experienced a paid trip into orbit.

Whatever the word, the experience of entering space is transformative for anyone who ventures there. International boundaries are almost invisible, Earthly disputes and politics shrink in importance, and a truer perspective emerges on human preoccupations and the fragility of our home planet. Returning astronauts often speak of gaining a new outlook not just of the Earth, but on life, and many others now have the potential to be similarly inspired.

Chapter One

Space voyagers

─────●──────────────────────

The dashing space traveller existed in fiction long before the first astronauts and cosmonauts entered space in the 1960s. The role has now taken a firm hold in popular culture, and is regularly played out in books and films as the epitome of danger, bravery and solitude.

OPPOSITE Yuri Gagarin was the first true space traveller. This 2012 painting by Alexey Akindinov, titled *Gagarin's Breakfast*, reflects the artist's imagination about what the cosmonaut might have said and done: "In orbit, I lay a space table!" Gagarin's own humorous geniality and the fractal infinity of space are evoked as he breakfasts with a samovar and *baranki* bread rolls, cruising in a ship of the imagination past a blue Earth and crescent Moon.

The origins of the term 'astronaut' can be traced back into early 20th century fiction. However, its widespread public use only started in 1959, when the US space agency NASA (National Aeronautics and Space Administration) chose the term to describe the space flyers it was about to recruit for its Mercury project, designed to put a man into space.

Astronaut literally means *star navigator*, from the Greek *ástron* (στρον), meaning *star*, and *nautes* (νατης), the Greek word for *sailor*. It is likely an adaptation of the word aeronaut whose origins go back to the late 1700s in French as *aéronaute*, used to describe the first balloonists, the Montgolfier brothers. The similar term aquanaut, meaning an undersea explorer, dates from the 1880s.

Astronaut, aeronaut and aquanaut, in turn, all seem to reflect ancient Greek literature. The compound word Argonaut was used to describe the sailors of the Greek ship *Argo* who searched with their leader Jason for the mythical Golden Fleece in Homer's epic, the *Odyssey*, composed in the 8th century BC. The *Argo* was named after its builder, Argus, who also made the voyage and was therefore himself an Argonaut.

The astronaut in fiction

The space traveller appeared in fiction before the term astronaut arose. The work of the French novelist, poet and playwright Jules Verne is often recognised as marking the advent of science-based adventure fiction. His novel *De la Terre à la Lune (From the Earth to the Moon)* of 1865, and its sequel *Autour de la Lune (Around the Moon)* published five years later, describe a three-man voyage from Florida to the Moon, which ends in a splashdown in the Pacific Ocean. The books curiously anticipate the first Moon landing over a century later in their launch and landing points, and the fact that Verne's cannon was called the *Columbiad*, while the Apollo 11 command module was named *Columbia*. The crew of *Columbiad*'s projectile are referred to simply as 'travellers'.

The word astronaut certainly emerged first in fiction. In 1880, the Manchester-born journalist and author Percy Greg (1836–89) wrote a lively novel titled *Across the Zodiac: The Story of a Wrecked Record*. It relates an account of a voyage by spaceship to Mars in 1830, powered by a fanciful anti-gravity substance, *apergy*. Considerable scientific accuracy is reflected in the author's description – the 100ft (30m) long vessel needs insulated walls 3ft (1m) thick to resist the outward pressure of internal air and the cold of outer space, and the flight takes advantage of the two-yearly proximity of Mars to Earth at opposition in order to minimise journey time.

In shape my *Astronaut* somewhat resembles the form of an antique Dutch East-Indiaman, being widest and longest in a plane

ABOVE In Greek mythology, Jason and the Argonauts set out to seek the Golden Fleece. This scene from their adventures is by English painter Herbert James Draper (1863–1920).

RIGHT Jules Verne's story *From the Earth to the Moon* imagines space voyagers making the journey in a projectile shot from a giant cannon located in Florida.

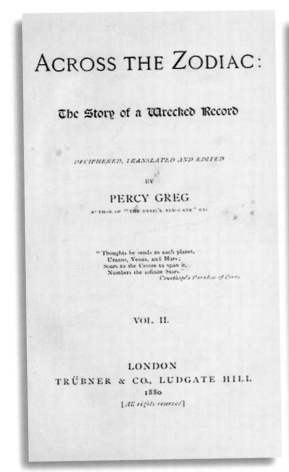

ACROSS THE ZODIAC:

The Story of a Wrecked Record

DECIPHERED, TRANSLATED AND EDITED

BY

PERCY GREG

AUTHOR OF "THE DEVIL'S ADVOCATE" &c.

" Thoughts he sends to each planet,
Uranus, Venus, and Mars;
Soars to the Centre to span it,
Numbers the infinite Stars."
Courthope's Paradise of Birds.

VOL. II.

LONDON
TRÜBNER & CO., LUDGATE HILL.
1880
[All rights reserved]

The return.

FAR LEFT English author Percy Greg's story *Across the Zodiac* of 1880 features a spaceship called *Astronaut* which is powered by an anti-gravity substance.

LEFT John Jacob Astor's novel *A Journey in Other Worlds* recounts a voyage to the outer planets in the year 2000 in a basket-like spaceship named *Callisto*. Illustration by Daniel Beard, 1894.

equidistant from floor and ceiling, the sides sloping outwards from the floor and again inwards towards the roof. The deck and keel, however, were absolutely flat, and each one hundred feet in length and fifty in breadth, the height of the vessel being about twenty feet…. Taking leave, then, of the two friends who had thus far assisted me, I entered the *Astronaut* on the 1st August about 4.30pm.

Across the Zodiac: The Story of a Wrecked Record,
by Percy Greg, Trubner, 1880.

After breakfasting on coffee, a weightless boiled egg eaten with difficulty from an egg-cup, and a small cigar, the unnamed narrator navigates to Mars and lands safely at sunset on a mist-shrouded landscape, to begin his planetary adventures. These include encounters with 'Martials', the local inhabitants whom we would today call Martians.

Greg certainly had some knowledge of Greek, because the book's fictional editor suggests that the name of the anti-gravity substance *apergy* derives from a combination of the Greek for work, *epos* (ργοζ) and *en-ergy*. This strongly implies Greg would have been familiar with Homer's *Odyssey* and the Argonauts, and establishes a link between the two words astronaut and Argonaut. Although the hero of the book would in modern parlance be known as an astronaut, the author uses the word only as the name of his spaceship, and not to refer to the traveller as a person. Greg's imaginative contribution has earned him a Martian crater named after him on Promethei Terra in the planet's southern hemisphere.

The useful material *apergy* re-appears in *A Journey in Other Worlds* (1894) by John Jacob Astor IV. This science fiction novel describes a voyage in the year 2000 to the planets Saturn and Jupiter, in a vessel called *Callisto*. Astor was better known as a wealthy American businessman who drowned in the sinking of the RMS *Titanic*.

Les
Navigateurs
de
l'Infini

ROSNY AÎNÉ

Préface de Jacques Bergier

ℛ

ÉDITIONS RENCONTRE

ABOVE Belgian writer J.-H. Rosny the elder may have been the first to use the word *astronaute*, which appears in his 1925 novel, *Les Navigateurs de l'Infini*.
(David Woods)

In French, the adjective *astronautique* (astronautic) was used in 1925 by J.-H. Rosny aîné (Rosny elder), the pen name of Belgian science fiction writer Joseph Henri Honoré Boex (1856–1940) on the model of the by then well-established word *aéronautique*, deriving from balloon flight. The word appears in Rosny's most famous novel, *Les Navigateurs de l'Infini* (*The Navigators of the Infinite*). The second part of his two-part story is entitled *Les Astronautes*, apparently using the word to refer to the role for the first time, but not describing a specific individual as such.

Rosny's book is set shortly after the exploration of the Moon, and describes travel to Mars in a spaceship named *Stellarium*, powered by artificial gravity, where the human explorers meet and consort with the local inhabitants. He had already written a much earlier novel set in prehistoric times, *Les Xipéhuz*, published in 1887, which is now regarded as one of the earliest stories in the science fiction genre.

French writer Théo Varlet, born in 1878 in Lille, was excited by *Les Xipéhuz* in his youth,

and went on to study and write about biology and astronomy. His 1930 novel *La Grande Panne* (*The Great Failure*) is significant for making the first use in French, and probably any language, of the term *astronaute* as it applies to a specific person. Varlet describes a young female astronaut, Aurore, who passes beyond the stratosphere and journeys to the Moon, where she discovers a type of organic dust.

It appears that credit for the first use of the term astronaut in the English language, to refer to an occupation or role as opposed to the name of a spacecraft, should go to the little-known American author Neil R. Jones (1909–88). Jones was a New York writer who produced several ground-breaking stories and one novel. His first tale, *The Death's Head Meteor*, was published in the science fiction magazine *Air Wonder Stories* in 1930, and describes how a Meteorological Bureau sets out to 'send space flyers [spaceships] to chart the open spaces for scientific research'.

The character Jan Trenton is introduced along with his novel occupation, and a description of the small spacecraft he uses to explore and retrieve the resources of meteoroids, which the author erroneously calls 'meteors', and which would nowadays be regarded as small asteroids:

The young astronaut approached his tiny space flyer. It was shaped like an egg, except that it was more elongated, and the two ends tapered down to blunt points instead of being rounded. It was mounted upon four revolving metal spheres set into its keel instead of wheels as landing gear. It was especially adapted for the use of exploring meteors, for all sides were studded with grapples and jointed drills as well as claw-like iron rods. These latter, which were also jointed, were capable of acting in the capacity of fingers in grasping material and placing it into the receptacles which lined the sides of the little space car. All of the exterior apparatus was manipulated by mechanical control from within.
The Death's Head Meteor, by Neil R. Jones, in *Air Wonder Stories*, January 1930.

The tale relates Trenton's mission to collect drill samples of the asteroid material, not dissimilar from missions proposed for modern astronauts

RIGHT *La Grande Panne* by French novelist Théo Varlet used the term *astronaute* to refer to a specific person in 1930.

feuilleton de la "Presse"
LA GRANDE PANNE
PAR
Théo Varlet
REPRODUCTION AUTORISÉE PAR LA SOCIÉTÉ DES GENS DE LETTRES

The First Men in the Moon

By

H. G. Wells

Author of "Tales of Space and Time,"
"Love and Mr. Lewisham,"
and "Anticipations"

"Three thousand stadia from the earth to the moon. . . . Marvel not, my comrade, if I appear talking to you on super-terrestrial and aerial topics. The long and the short of the matter is that I am running over the order of a Journey I have lately made."—LUCIAN's *Icaromenippus*

London
George Newnes, Limited
Southampton Street, Strand
1901

FAR LEFT American author Neil R. Jones described the work of an astronaut for the first time in the English language in 1930, as he travels the solar system sampling asteroids. *(Air Wonder Stories)*

LEFT H.G. Wells explained human locomotion on the Moon with remarkable accuracy in 1901.

in the near future. After a narrow escape from a grazing encounter with the atmosphere of Mars, the story ends with the reflective thought: 'Such is the life of an astronaut.'

Although Jones's somewhat uninspired writing style is now largely forgotten, the prolific science fiction author Isaac Asimov read his works when young, and acknowledges Jones's cyborgs, the Zoromes, as the 'spiritual ancestor' of his positronic robot series, and the source of his idea of benevolent robots. A machine of that description, NASA's dexterous humanoid Robonaut2, currently assists astronauts on the International Space Station.

Jones himself drew inspiration from earlier works, including H.G. Wells's *The First Men in the Moon*. Although he never used the term astronaut, British writer Wells displayed remarkably accurate knowledge of the lunar environment for someone writing in 1901:

I had forgotten that on the moon, with only an eighth part of the earth's mass and quarter of its diameter, my weight was barely a sixth of what it was on earth. ... I forgot once more that we were on the moon. The thrust of my foot that I made in striding would have carried me a yard on earth; on the moon it carried me six…

The First Men in the Moon, by H.G. Wells, George Newnes, 1901.

A few years after Jones's story appeared, the *Bulletin of the British Interplanetary Society* carried a poem *The Astronaut* by science fiction writer Eric Frank Russell, which suggests the term was now in wider circulation, at least among the early enthusiasts of spaceflight. The astronaut of the title speculates about flying by rocket to find 'an oasis in star-spangled space' where he might meet aliens. The hand-printed *Bulletin* was the Society's first publication, edited by science fiction author Arthur C. Clarke, and the poem appeared in the November 1934 edition.

Buck Rogers is a fictional spaceman created by Philip Nowlan in 1928 for the first science fiction magazine, *Amazing Stories*. He has been continually re-invented over the years for comic

BELOW Science fiction writer Eric Frank Russell's poem was published in the *Bulletin of the British Interplanetary Society* in 1934, suggesting that the term *astronaut* was becoming more widely used. *(British Interplanetary Society)*

Bulletin of the British Interplanetary Society. November, 1934.

THE ASTRONAUT
by E.F. Russell.
(Member-Liverpool)

Let me burst the Earth-bonds as a soul that is freed;
    Let me roar thru' the roof of the night
Like a man-conceived comet, competing in speed
    With the unsurpassed fleetness of light;
E'er onward, aye onward, 'til I've become one
    With the most distant dots in the dawn,
When each rumble of rockets marks many miles done
    And each planet a day newly born.
Let me pass thru' the aeons of light-years that bar
    All the knowledge of kin from their kind;
Let me meet them and greet them as one from afar
    Who has proved that the seeker shall find.
Let me quest an oasis in star-spangled space
    Where the sun and the nether-suns burn;
If I can find "Welcome!" on one alien face
    I care not if there be no return!
    +   +   +   +

strips, novels, movies, radio and television. In his adventures, the character encounters Martians and inhabitants of the asteroid belt, and travels the galaxy in search of long-lost human colonies.

Buck was still in action as late as the 1980s, having been transformed into the commander of *Ranger 3*, a fictional NASA craft that looked remarkably like the Space Shuttle. The shaping of the public's view of space exploration has been attributed to the Buck Rogers adventures, and their popularity stimulated a wider output of competitor stories, including *Flash Gordon* in 1934. Phil Nowlan's children would later tell Michael Collins of Apollo 11 that their father related the story of Buck Rogers's first flight to the Moon to them on 21 July 1934, almost precisely 35 years before the real landing.

Official use of the term astronaut

The official history of NASA's Johnson Space Center in Houston records that in 1959, when the United States and USSR were planning to launch humans into space, NASA Administrator T. Keith Glennan and his Deputy Administrator, Dr Hugh L. Dryden, discussed whether spacecraft crew members should be called *astronauts* or *cosmonauts*. Dryden preferred cosmonaut, on the basis that flights would occur in the *cosmos* (near space), while the *astro* prefix suggested flight to the stars. Most NASA Space Task Group members preferred astronaut, which soon became common usage as the preferred American term. When the Soviet Union launched the first man into space, Yuri Gagarin, they chose a term (космонавт) which anglicises to cosmonaut.

Today NASA applies the term astronaut to any crew member aboard a NASA spacecraft bound for Earth orbit or beyond. The agency also uses it as a title for those individuals selected to join its Astronaut Corps. The European Space Agency uses the same term for its astronauts, even before they have flown in space.

Spaceflight records are established by the FAI (Fédération Aéronautique Internationale), and its Sporting Code states that the term astronaut may apply both to crew members and to scientific personnel aboard the spacecraft who play an active part during the flight.

RIGHT Space hero Buck Rogers was invented by American writer Philip Nowlan in 1928 for *Amazing Stories*, the first science fiction magazine.

BELOW Hugh Dryden (left) and T. Keith Glennan (right) of NASA debated whether to call the first American space voyagers astronauts or cosmonauts. They are pictured here in 1958 with US President Dwight Eisenhower, Glennan being commissioned as NASA Administrator and Dryden as his deputy.

Although astronaut became the American term for the individual space flyer in 1959, the field of astronautics was already well recognised. The annual International Astronautical Congress was established in 1950, and the next year the International Astronautical Federation was founded in Paris as an organisation for international space co-operation.

Astronauts in movies

The first realistic depiction of human spaceflight in cinematic history was German director Fritz Lang's 1929 movie *Frau im Mond (Woman in the Moon)*. The plot centres around a flight to the Moon to seek gold, with an implausible love triangle which disturbs relationships between the six crew members. In making the film, Lang was advised by rocket pioneer Hermann Oberth and spaceflight advocate and writer Willy Ley. The movie authentically displayed rocket travel for the first time, including the use of a multi-stage rocket. The control centre performed a countdown to the moment of lift-off, the first use of this now-common device. It was popular with Wernher von Braun, who painted the logo of the movie on his V-2 rockets during the Second World War, and would later design the Saturn V rocket for America's Moon program.

Rocketship X-M (*X-M* stands for Expedition Moon) is a black-and-white movie from 1950, filmed in a rush to capitalise on the publicity of the delayed, bigger budget *Destination Moon*. It built upon the excitement of post-war American experiments with captured German V-2 rockets, and used film of an actual launch. It had some entertaining implausibilities, such as the crew commencing a leisurely smoke-filled press conference a mere 16 minutes before lift-off, and the commander wearing a tie in orbit!

The crew's lunar expedition is thwarted when, due to a malfunction, they survive encounters with meteoroids and end up on Mars instead. The descent to Mars was prophetically realistic, showing a landscape of winding dry river beds and canyon walls, interspersed with craters, which was unknown until Mariner 9 imaged the planet from orbit in

LEFT The motto of the International Astronautical Federation reads: Astronautics for Peace and Human Development.

1972. Lloyd Bridges starred as the pilot, and would later reprise an astronaut role in the 1970s TV series *Battlestar Galactica*.

Producer George Pál's *Destination Moon* was released just a month after *Rocketship X-M*, and gave an overall more satisfactory impression of its subject. Filmed in colour almost two decades before the first Moon landing, it was extraordinarily realistic for its vintage. Real telescopic images of the Moon's surface were used to depict a rocket landing inside the crater Harpalus on Mare Frigoris. Launch, the effects of weightlessness and the

BELOW The movie *Rocketship X-M* (Expedition Moon) was rushed out in 1950 ahead of *Destination Moon*, a bigger budget production then in the making. *(Lippert Pictures)*

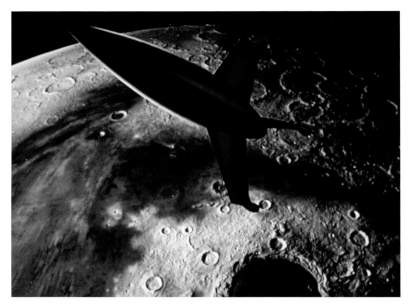

ABOVE George Pál's accomplished 1950 movie *Destination Moon* took advice from science writers and astronomers to achieve the most accurate depiction of spaceflight up to that time. The landing site in Harpalus crater touches the left side of the lower fin of the rocket. *(George Pál Productions)*

BELOW Movie director George Pál and early science fiction writer Percy Greg have craters named after them in Promethei Terra on Mars. *(NASA/W D Woods)*

ABOVE Coloured spacesuits in *Destination Moon* were to enable different astronauts to be distinguished in photographs – anticipating a problem experienced on the first two Apollo moon landings. *(George Pál Productions)*

one-sixth lunar gravity on the astronauts were realistically portrayed, thanks to the script-writing accuracy of science fiction author Robert Heinlein.

Realising that astronauts in identically coloured spacesuits would be indistinguishable from one another at a distance, the fictional project managers in the film decided to make each astronaut's suit a different colour. This cleverly anticipated a problem, unforeseen by real space mission planners, that would arise in lunar surface photographs taken by the crews of Apollo 11 and 12 in 1969. It was rectified for later missions by adding red bands to the commander's white spacesuit.

The time lag of radio communications to the Moon was also correctly depicted in the movie. Even the words of Dr Charles Cargraves, played by Warner Anderson, as he stepped on the Moon ("I take possession of this planet on behalf of and for the benefit of all mankind") anticipated what Apollo 11 commander Neil Armstrong would later read out to a listening world from the plaque on *Eagle*, the first craft to land on the Moon ("Here Men from the planet Earth first set foot upon the Moon, July 1969 AD. We came in peace for all mankind").

The landing site at Harpalus was chosen by astronomical artist Chesley Bonestell, who had earlier provided the ground-breaking illustrations for the 1949 book *The Conquest of Space* by Willy Ley. Harpalus has a relatively high lunar latitude, so the Earth would be low enough in the sky to appear

RIGHT An exaggerated lunar landscape greets the Moon voyagers in the 1964 movie of H.G. Wells's *The First Men in the Moon*.
(Columbia Pictures)

correctly positioned during camera shots. *Destination Moon* justifiably won the Academy Award for Visual Effects. In 1953, Pál also brought H.G. Wells's *War of the Worlds* to life in the movie of that name, but the space travellers were Martians coming to Earth. A Martian crater has been named after Pál, close to that of Percy Greg.

Wells's novel *The First Men in the Moon* was made into an entertaining movie in 1964. Supported by innovative special effects creator Ray Harryhausen, director Nathan Juran recounted the original tale of lunar voyagers claiming the Moon for Queen Victoria, but framed within the story of a modern United Nations lunar mission which discovers a tattered Union Jack on the surface. Wells has his own craters on both the Moon and Mars.

British science fiction writer Arthur C. Clarke and director Stanley Kubrick collaborated on the iconic movie *2001: A Space Odyssey*, released in 1968 with a tie-in novel. Astronaut Dave Bowman, played by Keir Dullea, pits his wits against the computer HAL, who is arguably the true star. The sequences filmed with models showing an orbiting space shuttle and station, a lander flying to the Moon, and a deep space mission to Jupiter involving spacewalks are extraordinarily convincing for the period and have rarely been bettered.

Based on a novel by Martin Caidin, the movie *Marooned* was released several weeks after the launch of Apollo 11 in 1969. Starring Gregory Peck and David Janssen, an Apollo spacecraft is stranded in orbit when its engine malfunctions, and a Russian vehicle comes to the rescue. NASA assisted in ensuring the realism of its spacecraft, and the movie won an Academy Award for Visual Effects.

A special place in Russian space history is reserved for the popular movie *White Sun of the Desert*, directed by Vladimir Motyl. This 1970 cult adventure is watched by

ABOVE Stanley Kubrik's 1968 movie *2001: A Space Odyssey*, with Keir Dullea playing astronaut Dave Bowman, has rarely been bettered.
(Metro-Goldwyn-Mayer)

LEFT *Marooned,* starring Gregory Peck and featuring an Apollo spacecraft, was released shortly after the first Moon landing flight, Apollo 11.
(Columbia Pictures)

ABOVE A few days before launch, Russian cosmonauts observe the tradition of watching the cult Soviet adventure movie, *White Sun of the Desert*. (Mosfilm)

BELOW In *Capricorn One*, a Mars flight is abandoned but political pressure ensures the hoax of its progress is deceptively enacted in a film studio. (Warner Bros.)

cosmonauts before their space launches as a good luck ritual. The names of the wives of one of the lead characters have been officially assigned to craters on the planet Venus.

In Peter Hymans's 1977 thriller *Capricorn One*, a manned Mars flight is cancelled shortly prior to launch and the crew are persuaded to act out the mission for television as if it had gone ahead. Its theme appears to have been one of the drivers behind the demonstrably untrue Moon landing hoax theory.

Despite a few historical inaccuracies, The Ladd Company's *The Right Stuff* (1983) from director Philip Kaufman is an impressive historical movie about the test pilots who broke the sound barrier and made the first American spaceflights in the Mercury program of the

ABOVE The dramatic stories of America's first Mercury astronauts are chronicled in *The Right Stuff*, released in 1983. (The Ladd Company)

1960s. Based on a book by Tom Wolfe, the three-hour epic portrays many dramatic and humorous anecdotes of the early Space Age. It starred Sam Shepard, Ed Harris and Scott

LEFT Tom Hanks, the ultimate astronaut actor, plays commander Jim Lovell in the movie of his near-disastrous Moon flight, *Apollo 13*. Here, he leads his crew out to board their spacecraft.
(*Universal Pictures*)

Glenn, and despite being a commercial failure, won four Academy Awards.

Director Ron Howard's *Apollo 13* is in the top rank of realistic space movies. It took a quarter of a century to turn the true-life drama of the greatest cliff-hanger in space history into a movie, adapting the book *Lost Moon* by mission commander Jim Lovell. Tom Hanks, the ultimate astronaut actor, plays the part of Lovell making his second trip to the vicinity of the Moon and hoping for a landing which he never achieves because his spacecraft is badly damaged by an explosion. The movie was made in 1995 before the era of widespread computer-generated imagery, so all the impressive scenes of rocket launches and spacecraft manoeuvres were filmed using detailed models, and some space scenes were filmed in actual weightlessness inside a NASA training aircraft. At the end of the movie the real Jim Lovell, by then having the US Navy rank of captain, makes a cameo appearance as the captain of the aircraft carrier USS *Iwo Jima* which lifted the Apollo 13 capsule and its crew from the Pacific Ocean.

Tom Hanks's enthusiasm for Apollo saw him collaborate again with Howard to produce an excellent 12-part television series entitled *From the Earth to the Moon*, released by HBO in 1998. It was based on *A Man on the Moon*, the unrivalled account of the Apollo program by Andrew Chaikin published in 1994. Each episode presents a different perspective, looking at the human side of the astronauts and their families as well as the accomplishments of their extraordinary missions.

On the 40th anniversary of the mission,

the flight of Apollo 11 and the human stories around it were recreated for television in the 90-minute drama *Moonshot*, directed by Richard Dale of Dangerous Films, London, with a tie-in book by Dan Parry. It too was notable for its attention to detail. Neil Armstrong

LEFT The television mini-series *From the Earth to the Moon* recreates the drama of the American space program through to the final lunar landing in 1972. (*HBO*)

ABOVE Television drama *Moonshot* relived the first Moon landing on its 40th anniversary. (Dangerous Films, 2009)

was played by Shakespearian actor Daniel Lapaine, his fellow moonwalker Buzz Aldrin was portrayed by James Marsters, and Andrew Lincoln had the part of Michael Collins, who remained in lunar orbit.

As the 50th anniversary of Gagarin's flight approached in 2011, BBC film producer

Chris Riley had an inspirational idea. Gagarin had no camera, so there was no on-board photographic record of that brief historic flight. The orbit of the ISS, however, lies at a somewhat similar inclination to that of Vostok. This meant that every so often, the ISS would more or less trace out Gagarin's ground track, passing over the same places on Earth. To obtain the correct lighting conditions, the station also had to pass over the Baikonur Cosmodrome at the same time of day, and in the same direction, that Gagarin was launched.

Italian astronaut Paolo Nespoli was then aboard the ISS, and armed with precise timing instructions, he managed to film the entirety of Gagarin's orbital track over a period of a few weeks. *First Orbit* is a captivating impression of the journey of Gagarin and the ISS around the Earth. It was shown across the globe on Yuri's Night, 12 April 2011, and so far has been viewed more than four million times on the internet.

RIGHT *First Orbit* was filmed from the International Space Station, recreating the view along the path traced out by Yuri Gagarin in his craft Vostok, during the first human spaceflight. (Christopher Riley/ The Attic Room/ Paolo Nespoli)

first orbit

a free film to download & share, created to celebrate the first 50 years of human spaceflight.

LIFTS OFF 12th APRIL 2011
www.firstorbit.org

RIGHT Sandra Bullock as astronaut Dr Ryan Stone floats out of her spacesuit in the movie *Gravity*, 2013. (Warner Bros.)

However, the movie that gives the most realistic impression of the space environment and the Earth from orbit is undoubtedly the 2013 release, *Gravity*, from producer-director Alfonso Cuarón. It stars George Clooney and Sandra Bullock as spacewalking Space Shuttle astronauts who are repairing the Hubble Space Telescope, when a cascade of deadly manmade space debris arrives. She escapes from one damaged craft to another and returns in a Chinese Shenzhou capsule. Critics take issue with several technological impossibilities, such as the orbital dynamics required to accomplish the space manoeuvres depicted, and Clooney wasting jet backpack fuel in aimless meanderings. However, the realistic sensations of motion, the captivating views of Earth and the imagery of spacecraft and spacesuits are unrivalled.

For meticulous accuracy in calculating technical and scientific parameters around a space mission, *The Martian*, both the 2015 movie starring Matt Damon, directed by Ridley Scott, and the original book by Andy Weir (2011), is hard to outdo. It relates the story of an astronaut stranded on Mars and his epic trek across the planet to facilitate his escape.

Doubtless there will be many more astronauts figuring in books, cinema, television and art in the future, and the output to date suggests the role has taken a firm grip of public consciousness and popular culture.

Factual space achievements have inspired an annual commemoration of space exploration on 12 April, the International Day of Human Space Flight, which is known as Cosmonautics Day in Russia. The festival of Yuri's Night is held in many countries and online, to honour Yuri Gagarin, who in 1961 pioneered human exploration of the cosmos in the world's first spacecraft, Vostok.

ABOVE Matt Damon fights for his life on a hostile planet in *The Martian*. *(20th Century Fox)*

LEFT Yuri's Night, held on 12 April, is the annual commemoration of the first human spaceflight.

FAR RIGHT Alan
Shepard, the first
American in space,
made a short sub-
orbital flight in 1961.
(NASA)

ASTRONAUT OR COSMONAUT?

The first human in space was called a
cosmonaut, and the second was an astronaut.
Both terms are still in use today. Does the
difference in terminology matter?

The two names reflect different traditions and
languages. Kosmos (космос in Cyrillic script) is
the Russian word for space, and it also exists in
English as cosmos, with the same meaning. In
Greek, ástron (στρον) is star and nautes (νατης)
is sailor.

The term *spaceman* was used in the early
days of the Space Age but has fallen out of use,
and its gender bias is no longer appropriate in
an era when both sexes regularly fly into space,
and women have commanded the US Space
Shuttle and the ISS.

For the first 17 years of the Space Age, the
only space voyagers were American astronauts
and Russian cosmonauts (космонавt in Cyrillic).
The western press often referred to Russian
cosmonauts as astronauts, and vice versa.
Curiously, during the first Moon landing, the
Soviet newspaper *Pravda* described Apollo
11 commander Neil Armstrong, using the
terminology of the discredited Russian royalty,
as the 'Tsar of the Ship'.

Then came the third nation to have a citizen
enter space, Czechoslovakia. When Vladimir
Remek flew with the Russians into orbit in 1978
as part of the USSR's Interkosmos program, he
was naturally also described as a cosmonaut.
Several other nations followed, including
Dumitru-Dorin Prunariu of Romania in 1982.
Today, being fluent in English, having served

on UN space committees and worked for the
European Space Agency, he is equally happy
with the title cosmonaut or astronaut.

Although astronauts and cosmonauts
floated into each other's spacecraft during
the joint Apollo-Soyuz mission in 1975, they
travelled into orbit and returned home in
their own vehicles. However, roll on a few
decades, and the distinction seems somehow
less relevant.

In 1994 Sergei Krikalev became the first
Russian to fly on the US Space Shuttle (STS-
60), wearing an orange US flight suit bearing
the Russian flag. Norman Thagard was the
first American to launch on a Russian craft
in 1995 on Soyuz TM-21 wearing a Russian
Sokol spacesuit with a US flag shoulder patch.
He might be regarded as the first American
cosmonaut.

When Jean-Loup Chrétien became the first
French citizen and first Western European in
space in 1982, a new term entered the fray.
He was called a *spationaut* (*spationaute* in
French, irrespective of the person's gender),
from the Latin word *spatium* for *space*.
However, current French space travellers
fly under the aegis of ESA, and are called
astronauts in English.

In 2003 China entered the ranks of space-
faring nations. Some non-Chinese news reports
referred to its crew members by the term
taikonaut, and the *Oxford English Dictionary*

BELOW Valentina
Tereshkova's 1963
flight on Vostok 6 was
a propaganda coup for
the USSR. The poster
proclaims: Glory
to the first woman
cosmonaut!

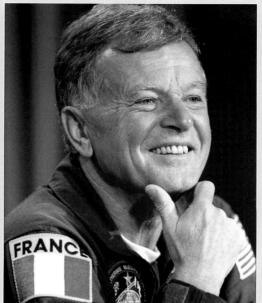

FAR LEFT Norman Thagard was the first US astronaut to launch on a Russian Soyuz spacecraft.

LEFT Jean-Loup Chrétien became the first French person in space in 1982, flying first with the Russians and then on the US Space Shuttle.

now recognises the word as a 'hybrid of the Chinese term *taikong* (space) and the Greek *naut* (sailor)'. However, taikonaut is more of an English nickname, and the proper Chinese term as used by official media when reporting on the nation's astronauts is *yuhangyuan* (航天员), which is Chinese for *space navigator*.

The Malay word for astronaut, *angkasawan,* was used to describe Malaysia's trainee astronauts, and Sheikh Muszaphar Shukor became the first Malaysian in space when he flew to the ISS on a Russian Soyuz in 2007. Not to be outdone, the Indian Space Research Organisation (ISRO) hopes to launch a crewed domestic spacecraft that would carry *vyomanauts*, coined from the Sanskrit word for *space* or *sky*. This is unlikely to be achieved before 2024 at the earliest. However, when Rakesh Sharma became the first Indian citizen to enter space in 1984, he flew with the Russians as a cosmonaut.

Space authorities are wisely resisting this proliferation of names, all meaning the same thing, and trying to agree to limit the terminology to the long-standing terms cosmonaut and astronaut.

A special case, however, arises with space tourism. The vast majority of these space travellers, and in particular those who make sub-orbital flights, will not receive the years of rigorous training imposed on professional astronauts. So NASA and Russia's Roskosmos

have agreed on the term *spaceflight participant* to describe people who travel into space, but are not professional astronauts.

The international professional and educational organisation for individuals who have orbited the Earth avoids even the cosmonaut/astronaut dilemma by calling itself the Association of Space Explorers. It was founded in 1985 and this once-exclusive club is growing rapidly, having over 400 flown astronauts and cosmonauts from 37 nations as members. It holds its annual congress every autumn in different countries. The qualification for membership is to have completed at least one orbit of the Earth, hence sub-orbital space tourists cannot expect to join.

ABOVE China joined the ranks of nations having an independent human spaceflight capability in 2003, when Yang Liwei flew into orbit on a Shenzhou spacecraft.

Chapter Two

Roles and missions

Modern astronauts are no longer the all-male test pilots they once were, and crews now require widely varied professions and specialisms of both genders. Mission destinations accomplished so far include sub-orbital flight, Earth orbit, Moon orbit and lunar surface. Plans are now being made for deep space locations such as the stable Lagrange points, asteroids and the planet Mars. Future astronauts may land on the moons of Mars and the outer planets.

OPPOSITE In this visualisation by artist Lucy West, three astronauts explore the cold, dimly lit surface of Neptune's large moon, Triton. In the background, a geyser of nitrogen is spouting, an example of the cryovolcanism observed there by the Voyager 2 space probe. *(Lucy West)*

While the initial cadres of cosmonauts and astronauts of the Soviet Union and USA were all culled from the ranks of the elite jet pilots, as were the more recent first *yuhangyuan* of China, the intake of personnel of different backgrounds has now widened considerably. The main reason for this change has been the advent of larger craft and space stations with scope to accommodate personnel who do not need to be able to pilot the vehicle, leaving that task to experienced specialists.

Roles

Spaceflight no longer simply requires a fighter pilot to steer a spacecraft into the sky and fly it in orbit. Space offers a wide variety of career opportunities for astronauts with occupations that include medical doctors, astronomers, physicists, geologists, journalists, teachers, politicians and business employees sent up to operate their company's equipment. This can only be expected to widen, once space tourism opens up.

The primary and essential role is a mission commander who is also a pilot of the spacecraft. He or she is the leader of the crew and will either fly the craft themselves, or oversee its automatic flight, usually in a combination of both active and inactive modes. However, they have to be capable of flying all stages of the mission in the event of anomalies. They would normally have a co-pilot or flight engineer on board.

Even if the spacecraft's standard flight mode is designed to be highly automated, as it is with the Russian Soyuz and the newest American vehicles, a human must be able to intervene on occasion, such as to complete a docking after the automatic systems have gone awry, or to take over from the computer to steer a lunar lander to a safe touchdown.

The International Space Station brings another layer of complexity to roles, because the commander of the ferry craft which brings a crew up from Earth and returns them home will not necessarily be the commander of the station. The command of the ISS mainly rotates between Russia and the USA, as the main funders, but other nationalities have commanded from time to time, including Belgium (as part of ESA), Canada and Japan. Hence, a Soyuz taxi that delivers new crew members to the ISS may have the ISS Commander among its crew, but for the journeys to and fro that person will be under the command of a Russian cosmonaut, the Soyuz Commander.

Gender is less of a barrier to these commander roles than it once was, and several women have commanded the US Space Shuttle and the ISS. Eileen Collins was

Sub-orbital flight

In the accepted modern definition of sub-orbital flight, the astronaut enters space, flying above the Kármán Line (see The Boundary of space on page 124) at an altitude of 62 miles (100km).

The first astronaut to undertake a sub-orbital flight was Alan Shepard in his capsule Freedom 7, during the Mercury-Redstone 3 flight, in May 1961. He became the first American in space, as he launched from Cape Canaveral in Florida, soared to an altitude of 116 miles (187km) and splashed down 15 minutes later in the Atlantic Ocean, having travelled just 302 miles (486km) down range. The achievement was overshadowed by Yuri Gagarin achieving the far greater feat of orbital flight just three weeks earlier aboard Vostok 1.

Virgil 'Gus' Grissom repeated the exercise in *Liberty Bell 7*, reaching higher at 118 miles (190km), and after splashdown was lucky to escape alive from his sinking spacecraft due to a faulty hatch jettison. The *Liberty Bell* capsule was finally discovered and retrieved from the bottom of the ocean in 1999.

Joe Walker was belatedly recognised as the world's third sub-orbital voyager in 1963. In July and August of that year, just after the conclusion of the Mercury program which put six American astronauts into space in

ABOVE America's first astronaut Alan Shepard and his Mercury capsule *Freedom 7* on the deck of the USS *Lake Champlain* after his 15-minute sub-orbital spaceflight. He travelled 302 miles (486km) from Cape Canaveral, peaking at 116 miles (187km) on 5 May 1961, three weeks after Yuri Gagarin's orbital flight in Vostok. *(NASA)*

BELOW The US Army Ballistic Missile Agency developed the Redstone rocket and this plan for Project Mercury's sub-orbital flights, dated May 1959. The flight arc was to be 190 nautical miles (219 miles, 352km) long, peaking at 125 miles (201km) altitude, with splashdown near Bermuda. Had it not been for repeated delays, Alan Shepard would have been in space before Gagarin.

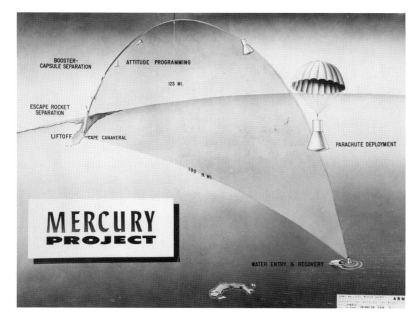

LEFT Joe Walker made two sub-orbital flights in the X-15 aircraft which were only officially recognised as spaceflights after his death. He was awarded his astronaut wings posthumously in 2005. *(NASA)*

rocket-launched capsules, Walker twice flew the X-15 well above the Kármán Line to 67 miles (108km) altitude. This in effect made the X-15, previously regarded as a rocket-powered aircraft, a proper spaceplane.

It was a long time before Walker got the recognition he deserved. He died in an air crash in 1966, but it took until the 1990s for his achievement to be recorded in space annals, and he finally received his astronaut wings from NASA posthumously in 2005. This required some re-ordering of spaceflight records, since as well as becoming the USA's seventh astronaut (previously regarded as John Young on Gemini 3), Walker is now accepted as the first US civilian astronaut and the first person to make two spaceflights (previously Virgil Grissom).

In April 1975 two Soviet cosmonauts, Vasili Lazarev and Oleg Makarov, launched in a Soyuz vehicle from Baikonur on a mission to the Salyut 4 space station. When out of the atmosphere, at

RIGHT X-15 Flight Profile. Take-off was from Wendover Air Force Base in Utah, where an NB-52 carrier aircraft took the X-15 to its launch altitude of 45,000ft (13.7km). After release, it rocketed out of the atmosphere, coasted to a peak altitude of up to 67 miles (108km) over Nevada, went through re-entry and glided down to an unpowered landing at Edwards AFB in California.

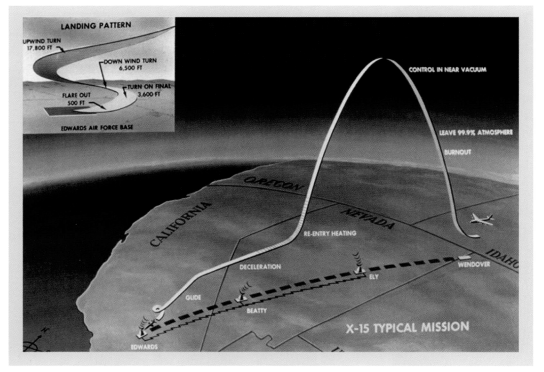

rocket has to rise out of the atmosphere and simultaneously curve over to travel parallel to the curving surface of the Earth, and reach a precise point in space far from the launch site, at a precise speed. Without sufficient velocity, it will fall back to Earth on a sub-orbital track and re-enter the atmosphere; too much speed, and it will soar off on a long elliptical path that will complicate any plans to return to Earth.

As with sounding rockets and unmanned satellites, so with crewed spacecraft. Orbiting the Earth is by far the greater technological accomplishment, compared with sub-orbital flight, requiring more complex launchers and spacecraft as well as much greater cost.

Early manned orbital flights involved single craft remaining in space for increasingly longer times. Next were joint flights of two and even three spacecraft. A major technical objective was complex manoeuvres to perform rendezvous and docking. Spacewalking cosmonauts and astronauts exited their vehicles for what became known as Extra Vehicular Activity (EVA).

A step change in capability came in 1981 with the reusable US Space Shuttle, which could carry considerable cargo into orbit, including laboratories for science experiments, deploy commercial and scientific satellites

ABOVE Gagarin's Vostok spacecraft, seen here in a computer-generated image, carried ten days' provisions in case the retro-rocket mis-fired and he had to wait until atmospheric drag brought the craft down to Earth.

BELOW The US Space Shuttle *Atlantis* on STS-135 makes the final flight of the program, passing over the Bahamas prior to docking with the International Space Station. Its cargo bay carries the Italian-built *Raffaello* multipurpose logistics module which is packed with supplies and spare parts. *(NASA)*

ABOVE Vapour trails spiral off the wing tips of *Atlantis* on mission STS-135 as it lands at the Kennedy Space Center on 21 July 2011, bringing the Shuttle program to an end. *(NASA/Sandra Joseph and Kevin O'Connell)*

BELOW Against the blackness of space and the thin line of Earth's atmosphere, the ISS is seen from Space Shuttle *Discovery* as the two craft slowly separate at the end of mission STS-119 in 2009. *(NASA)*

and send probes into deep space. It landed horizontally on a runway like an aircraft.

With the advent of space stations upon the launch of the USSR's Salyut in 1971, the basic spacecraft became primarily a taxi to transport crew to a more commodious facility, where they could spend longer in space and conduct increasingly sophisticated experiments, then bring them home again. Following the end of the US Moon landing program, three Apollo spacecraft conveyed crews to the Skylab space station and another conducted a space docking with Russia's Soyuz 19.

The Russian Soyuz fulfilled the same role in transit to and from several Salyut stations, Mir, and now the International Space Station. China's Shenzhou craft transports crews to its own Tiangong space station.

Further plans exist for new space stations owned by China and Russia, and commercial stations that are being designed to receive tourist astronauts and to host scientific experiments. For instance, Bigelow Aerospace already has an inflatable module attached to the ISS to test its designs for larger structures.

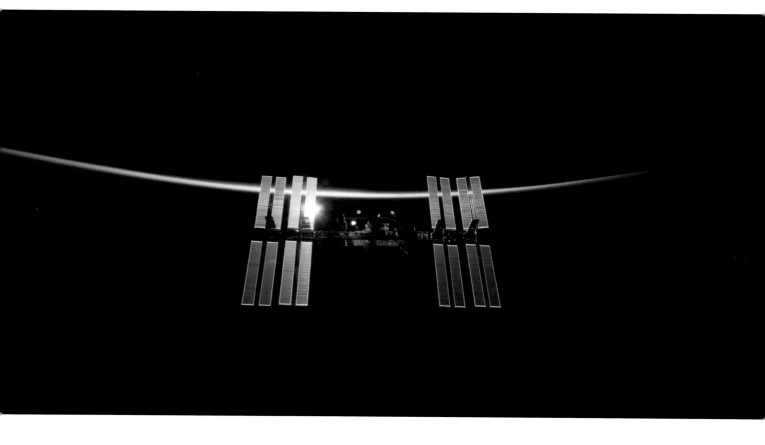

RIGHT Space Shuttle *Endeavour* on STS-61 was the first repair mission to the Hubble Space Telescope, in 1993. Astronaut Story Musgrave is on the end of the Remote Manipulator arm, about to install protective covers on the magnetometers. Jeffrey Hoffman is working inside the Shuttle payload bay. *(NASA)*

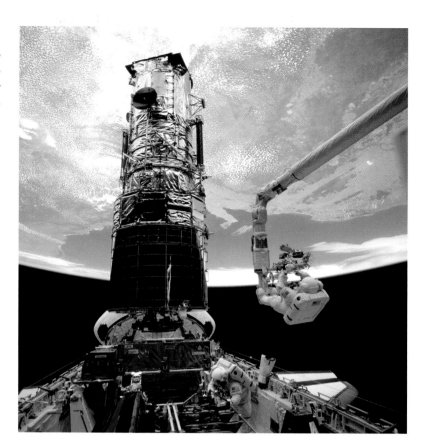

In 2020 it intends to launch a 45ft (14m) long by 22ft (6.7m) diameter inflatable habitat, designated B330 on account of its internal volume of 330m³.

Since Earth orbit is likely to remain the main destination for astronauts for the foreseeable future, it is the obvious location for research stations. It can also act as a stepping stone to other points in the solar system. For instance, all the manned lunar flights of the 1960s and '70s tarried briefly in Earth orbit for the crew to check out their craft before progressing to the next stage of their lunar voyage. Once sub-orbital space tourism is established, orbital tourism is sure to follow.

The region of space known as low Earth orbit (LEO) starts at an altitude of about 100 miles (161km) and extends to the somewhat arbitrary upper limit of 1,200 miles (2,000km). It is possible to orbit at a lower altitude, but air drag will cause rapid orbital decay and result in atmospheric re-entry.

The International Space Station orbits well inside the LEO zone, at an altitude of 250 miles (400km). Apart from flights to the Moon this altitude has rarely been exceeded by manned missions, with the most notable exceptions being Space Shuttle missions to deploy and later repair the Hubble Space Telescope. Shuttle *Discovery* on mission STS-82, commanded by Ken Bowersox, boosted the telescope's orbit to a peak of 385 miles (620km) in 1997.

Only two crews have ventured higher and that was, by modern standards, in a very basic spacecraft. Charles 'Pete' Conrad and Richard

RIGHT In 1966 Pete Conrad and Dick Gordon flew on Gemini XI to a record altitude for a non-lunar flight, reaching 854 miles (1,374km). They looked down on India and Sri Lanka over the Agena vehicle which they docked to. *(NASA)*

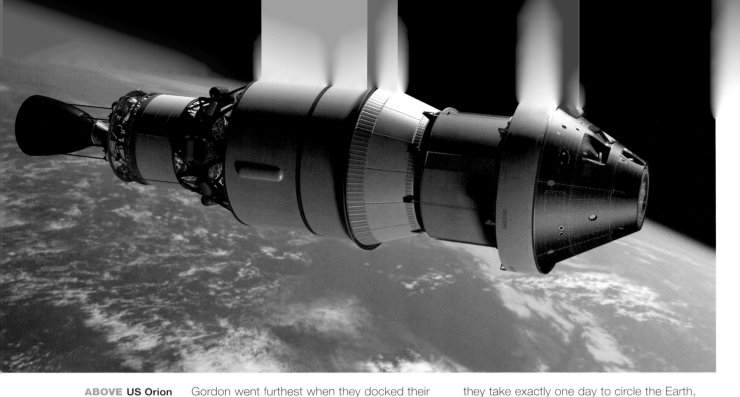

US Orion capsule and Delta IV upper stage during Exploration Flight Test 1 in 2014, when it travelled 3,600 miles (5,800km) from Earth, then returned to splash down in the Pacific Ocean off the coast of Mexico. In its first crewed flight, astronauts will use this vehicle to travel around the Moon.
(NASA)

RIGHT Joseph-Louis Lagrange (1736–1813) was an Italian mathematician and astronomer whose work on planetary orbits and the 'three-body problem' explained what are now known as the Lagrange points.

Gordon went furthest when they docked their Gemini XI craft to an Agena rocket, started up its engine, and soared to an altitude of 854 miles (1,374km), gaining an unprecedented view of the curvature of the Earth and taking some iconic photographs of Arabia and India.

High Earth orbit

There is a huge volume of space between low Earth orbit and the Moon, which orbits at an average distance of 239,000 miles (384,000km). It is mostly occupied by communications and observation satellites at an altitude of 26,200 miles (42,160km), where they take exactly one day to circle the Earth, and therefore appear from ground level to be stationary in the sky – perfect for aiming a TV satellite dish, or for continually monitoring the same face of the Earth below.

There was a plan to send an early manned Apollo mission out to 4,000 miles (6,450km) in 1969 but it was decided to fly to lunar distance instead. America's new Orion crew capsule had a successful uncrewed test flight to a peak altitude 3,600 miles (5,800 kilometres) in 2014.

Except for Apollo spacecraft in transit on the three-day voyages to and from the Moon, even today no crewed spacecraft has exceeded the 1966 Earth orbit altitude record set by Conrad and Gordon on Gemini XI.

Lagrange points

Named after Joseph-Louis Lagrange, the 18th–century Italian astronomer who discovered them, these are positions around two orbiting bodies where a spacecraft can hover in a stable position with minimal expenditure of fuel. There are five such balance points in a system in which one cosmic body orbits another, such as the Earth-Moon system, the Sun-Earth system, and any other planet which orbits the Sun. Two of these, designated L4 and L5, are naturally permanently stable, one being roughly 60° ahead of and the other 60° behind the orbiting body, and no fuel is required to hover. For several of the Sun-planet systems, these two points contain clouds of dust and small moonlets which linger permanently there. Several automatic observational satellites have been carefully manoeuvred to occupy two of the Sun-Earth Lagrange points and to fly through others.

However, it is the Earth-Moon Lagrange points, being closest to home, that are of most interest to the planners of human spaceflight. Some have been visited, but only by unmanned probes to date. Japan's Hiten spacecraft visited both the L4 and L5 points, en route to achieving lunar orbit at very low fuel consumption, compared with usual orbital transfers. NASA's ARTEMIS and China's Chang'e 5-T2 both visited L2. An orbit around L2 will be especially useful as the position for a relay satellite to transmit signals to Earth from future astronaut missions to the far side of the Moon. Lunar L5 was also the spot proposed for a well-publicised space colony by the American space activist Gerard K. O'Neill in the 1970s, whose location could readily receive construction materials from the Moon.

Both the US and Russia have assessed options for using Lagrange points as staging posts for future missions. For instance, NASA studied using the L1 point of the Earth-Moon system for a manned return to the Moon, and also as a precursor to crewed missions to Mars. A Gateway station at L1 would be reached by a Crew Transfer Vehicle bringing a crew of four from the International Space Station. Then, a crewed lunar lander would descend from the Gateway to the surface of the Moon. The mission also features a lunar habitat lander which would carry supplies and shelter to support a lunar crew for 30 days. Lunar L1 is an advantageous Gateway location because it offers access to the entire lunar globe, continuous sunlight for solar power, and

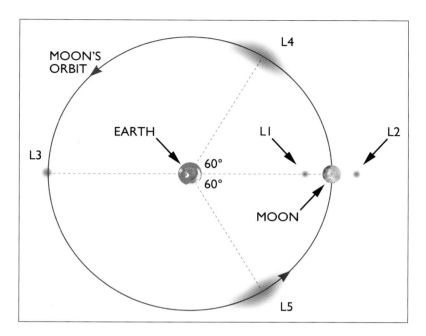

ABOVE Spacecraft can hover at the five Earth-Moon Lagrange points of gravitational balance, the larger areas (L4 and L5) being more stable. The entire system rotates around the Earth following the Moon's motion. L1, L4 and L5 could accommodate crew way-stations offering easy access to the lunar surface, while a satellite orbiting L2 could provide a telecommunications relay for operations on the far side of the Moon. *(David Woods)*

RIGHT NASA has evaluated a mission in which the Orion capsule would orbit the Earth-Moon L2 point. From here, astronauts could tele-operate surface robots on the far side of the Moon from Earth. The L2 point itself does not have line-of-sight to Earth, but a slow orbit around it could have. *(Lockheed Martin)*

uninterrupted communications with Earth. The thrust requirement for station-keeping is only 10m/s per annum, and there are return flight opportunities back to the space station every ten days.

Modern computational power has revealed that Lagrange points offer unique opportunities for highly efficient travel to other planets. The mathematical foundations are being laid for what has been termed the *Interplanetary Superhighway* which, by using the paths formed by the systems of Lagrange points throughout the solar system, will increase the pace of human space exploration and greatly reduce the costs.

The Moon

It is the most prominent object in our night sky, and far larger in relation to the Earth than any other moon in the solar system is to its parent planet. It stirred the curiosity, philosophy and speculation of the ancients, and beckoned imaginative travellers and their fanciful voyages for many centuries. Astronauts finally flew around the Moon's grey orb in 1968, and walked on its dusty surface in 1969. It remains the only celestial body to have been trod by human feet.

As memories of the Apollo program of the last century fade, so too does public awareness of the extraordinary achievements of the American astronauts who accomplished those lunar voyages. Between 1968 and 1972 the US undertook nine three-man expeditions to the vicinity of the Moon, and in six of those made surface landings. Three individuals made the lunar trip twice, hence a total of 24 men circumnavigated our nearest celestial neighbour. In the almost half a century since then, the furthest anyone has been from Earth is the Hubble Space Telescope, a mere tenth of 1 per cent of lunar distance.

All of these legendary astronauts who survive are now in their 80s. The name of Neil Armstrong will stand in history alongside

LEFT **The Earth and Moon viewed by the Deep Space Climate Observatory (DSCOVR) satellite positioned at the Sun-Earth L1 Lagrange point, 930,000 miles (1,500,000km) sunwards of Earth. The Earth-Moon L2 point could offer a somewhat similar perspective for a crewed expedition, enabling them to view simultaneously the Earth and the far side of the Moon, albeit from a closer distance. Visible are the Persian Gulf, Caspian Sea, Borneo and Australia, and on the Moon, Mare Moscoviense.** *(NASA)*

RIGHT Apollo 17 Command Module *America* in lunar orbit, December 1972, photographed from the Lunar Module *Challenger*. Its Scientific Instrument Module is on the right, with the docking probe at the bottom. *(NASA)*

Christopher Columbus, Ferdinand Magellan, Yuri Gagarin, Charles Lindbergh and the other icons of human terrestrial, aerial and cosmic exploration. The co-pilot who accompanied Armstrong on his historic descent to the first human landing and walk on the Moon, Dr Edwin (now 'Buzz') Aldrin, still has a high degree of public recognition. But the other moonwalkers are largely anonymous to popular history:

- Apollo 12, 1969: Charles 'Pete' Conrad, Alan Bean
- Apollo 14, 1971: Alan Shepard, Edgar Mitchell
- Apollo 15, 1971: David Scott, James Irwin
- Apollo 16, 1972: John Young, Charles Duke
- Apollo 17, 1972: Eugene Cernan, Harrison 'Jack' Schmitt.

Each mission had a third crew member who stayed in lunar orbit to operate the Apollo mothership. This role was essential to the safety of the landing crews and to their return to Earth, and is always acknowledged by the moonwalkers, though public recognition of their names is limited. For the record, they are

RIGHT The Moon has now been charted in much greater detail than this Lunar Topographic Orthophotomap derived from Apollo 15 survey data acquired in 1971. It shows the Apollo 15 landing site near Hadley Rille at 5,300m above lunar datum, with Mount Hadley to the north-east towering 13,500ft (4,100m) higher. *(NASA, 1975)*

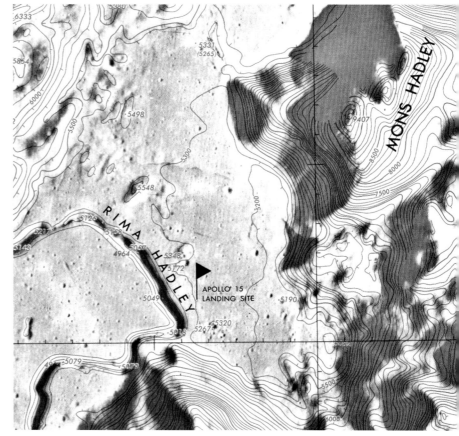

Michael Collins (11), Richard Gordon (12), Stu Roosa (14), Al Worden (15), Ken Mattingly (16) and Ron Evans (17).

The astronauts unfailingly also point out that their glamorous role was merely the apex of a gigantic pyramid comprising thousands of personnel who designed, built, tested and operated the spacecraft, rockets, launch sites, communications systems and other essential facilities.

Alan Bean, the fourth man on the Moon and an artist who paints lunar scenes in his retirement, acknowledges Armstrong's pre-eminence: "Neil is the only one of us that will be remembered in history."

The Cold War between the US and the Soviet Union undoubtedly accelerated the momentous achievement of the Moon landings, something that would surely have taken many decades longer to accomplish had technology progressed in a less politically motivated way. Now, as Apollo recedes into history, a more measured assessment of the role of the Moon in human exploration is possible.

Debates have raged in the intervening decades about the next destination for human spaceflight beyond the International Space Station. Some favour a return to the Moon, others a direct trip to Mars. However, we have hardly touched the Moon, never venturing further than a few kilometres from the six localised landing sites on a territory the size of Africa. Irrespective of fluctuating American opinions, it seems clear that both China and Russia would like to venture to the Moon or its vicinity, and this feat is much more readily achieved than an 18-month flight to and from Mars. Furthermore, recent research has identified potential resources on the Moon that might motivate a return.

Its proximity to Earth makes the Moon a suitable place to test hardware for medium-distance flights and surface operations, and to train crews in an environment that represents a realistic analogue for the greater perils of a long-duration trip to Mars.

Hence astronauts of the immediate future can expect to participate in several types of lunar mission, sometimes in combination – circumnavigation, orbit, landing and surface exploration.

In circumnavigation the spacecraft loops around the Moon without entering orbit, gets a close-up view of the surface, then re-appears on the opposite limb and heads directly back to Earth. The Russians achieved this in 1968 with their Zond craft, which were basically a stripped down Soyuz, carrying live animals but

BELOW US Orion spacecraft in lunar orbit, docked to one possible design for an advanced lunar landing craft.

no cosmonauts. They were ready to launch a crew on a repeat mission, but were narrowly beaten by the US Apollo 8 flight, so the project was quietly abandoned.

Jim Lovell, Fred Haise and Jack Swigert of Apollo 13 inadvertently became the only space crew to accomplish such a flight when their intended Moon landing was thwarted by an on-board explosion. As they passed 158 miles (254km) above the far side of the lunar surface on 15 April 1970, they set a record for the maximum distance humans have travelled from Earth, at 248,655 miles (400,171km), which still stands today.

Flights of this type offer an excellent testing and training opportunity for astronauts to fly to lunar distances without the additional complexity and fuel requirements of entering lunar orbit, and are likely to comprise part of the continuing test program for NASA's new Orion Multi-Purpose Crew Vehicle (MPCV). Such a flight is also envisaged by the company Space Adventures for one of the most extreme space tourism trips yet proposed, a loop around the Moon in a Russian Soyuz after a ten-day stay at the ISS.

Apollo 13 was a special case due to the accident that changed the flight plan and required a hasty return to Earth. The only other crewed flights to the Moon were eight Apollo missions, all of which entered retrograde lunar orbits, passing across the lunar disc from right to left as viewed from the Earth's northern hemisphere.

Starting with Apollo 8 in 1968, astronaut missions to lunar orbit were used to reconnoitre landing sites and to rehearse procedures prior to the first landing attempt. Nowadays, reconnaissance is done more economically by unmanned probes, and the more useful orbits for manned missions are thought to be polar, which can give access to the entire lunar surface over the course of a month, as the Moon rotates. This approach was successfully trialled by the unmanned Chinese lander, Chang'e 3 in 2013.

From such an orbit, a crewed lander has an opportunity to descend to either the North or South Pole on every orbital rotation. This offers great flexibility to target these key locations, which have been proven to hold useful resources of water ice, for future crewed landings.

Once on the lunar surface, a wide range of tasks will be required of the astronauts. They will have to investigate and photograph their surroundings, collect rock and soil samples, set up independently powered experiments packages and construct habitats. Sometimes they will conduct walking EVAs, and on longer trips will have access to electric-powered vehicles. On three of the Apollo missions the astronauts used a two-person unpressurised Lunar Roving Vehicle (LRV) that bore a resemblance to a dune buggy. Future expeditions will have access to a pressurised vehicle which can be driven in a shirt-sleeves environment, then parked for spacesuited moonwalks after exiting through an airlock.

Asteroids

Beyond the Moon, the closest celestial bodies to Earth are the so-called Near-Earth Asteroids (NEA). To be classified as such, they need to come closer than 31 million miles

BELOW Astronaut-geologist Jack Schmitt at the Apollo 17 landing site in the lunar valley of Taurus Littrow. Schmitt is opening the solar panels on the transmitter for the Surface Electrical Properties experiment. Because of the stiffness of his pressurised suit in the lunar vacuum, Schmitt can only reach down by putting his right leg back and leaning sideways. Lunar Rover LRV3 is on the right, its umbrella-like antenna pointing to Earth. The lunar lander *Challenger* sits in the distance with the smooth slopes of the South Massif beyond. *(NASA/Eugene Cernan/David Woods)*

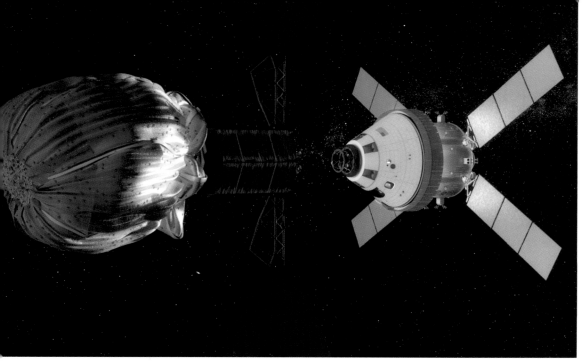

(50 million km) to the Earth's orbit around the Sun. Many, however, are much closer and some sail past our planet inside the orbit of the Moon, and occasionally even below the orbits of telecommunications satellites. About 15,000 have already been identified, and they vary in size from several metres up to 20 miles (30km) across.

About 1,000 NEA are half a mile (800m) in diameter, and the more accessible ones offer an excellent opportunity to test systems for crewed flight in deep space, rendezvous and surface sampling. This can be accomplished without venturing as far as the nearer planets, and without the difficulties posed by strong gravity and atmospheres. They are also fascinating science targets in their own right, and the danger that some may pose to Earth as potentially devastating impactors lends a certain relevance and urgency to investigating them.

Several more distant asteroids have already been visited close-up by unmanned spacecraft, including Eros and Itokawa, and many more

have been flown past and photographed, or imaged by radar from Earth. NASA's OSIRIS-REx mission is on its way to retrieve samples from the asteroid Bennu, which is about 500 yards (metres) across.

These are all helpful precursors for an eventual astronaut mission to an asteroid. The first may be NASA's Exploration Mission 2 of the Orion MPCV, where four astronauts could rendezvous with a very small asteroid that has been captured and placed in lunar orbit, perhaps as early as 2023. With larger asteroids, later missions have been outlined in which astronauts with jet-propelled backpacks would fly over from their spacecraft to land on the object, gather samples and perhaps prepare it for capture. This would be more akin to a weightless EVA than a moonwalk, because the extremely low gravity will make foot traction in

slippery dust almost impossible. A kick might send an astronaut off at escape velocity, and a spacecraft may orbit around it at walking speed.

Mars

Mars is the next obvious destination for human space exploration, the one most speculated about in science fiction, and probably the most benign environment for creating a human settlement. Many serious engineering studies have been undertaken since the 1960s, and a cavalcade of unmanned orbiters, landers and surface rovers has returned an extraordinary wealth of knowledge about the planet.

Its surface gravity is 38 per cent of the Earth's and over twice that of the Moon, making human mobility comfortable. Its tenuous atmosphere provides some protection from meteoroid impact, and will even support some wispy clouds high in the sky and other Earth-like atmospheric phenomena such as dust storms and small whirlwinds. At the equator, the midday summer Sun can lift temperatures

BELOW Martian sunset at Gale Crater imaged by the rover *Curiosity* in 2015. Fine dust in the thin Mars atmosphere scatters blue light forward to create a cool blue sunset. *(NASA/JPL-Caltech)*

above zero degrees Centigrade (32°F). It would be quite tolerable for humans, were it not for the lack of oxygen and the extremely low pressure which necessitate a spacesuit.

Mars is the long-term target for some national space programs, including NASA, which plans to use its Orion spacecraft and additional hardware such as a Deep Space Habitat module and a propulsion stage to reach Mars, orbit the planet, and return to Earth sometime in the 2030s. A landing mission is anticipated to follow shortly thereafter.

However, the ambitious goal of entrepreneur Elon Musk and his commercial spaceflight company SpaceX is to establish a colony on Mars. The company intends to send its robotic Dragon 2 capsule (also called Red Dragon, and very similar to the Crew Dragon to be used to deliver astronauts to the ISS) on a one-way test flight to Mars, possibly as early as 2018. It will descend by rocket-braking and demonstrate technologies for landing large payloads on the planet, including supplies and habitats for Martian explorers. Future plans of SpaceX

ABOVE Wind tunnel testing of a parachute for a Mars landing. In 2016, ESA's *Schiaparelli* probe employed this parachute of the 'disc-gap-band' type that has also been used for all NASA planetary entries. *(USAF Arnold Engineering Development Complex/ESA)*

BELOW SpaceX's Red Dragon capsule is visualised sitting on the surface of Mars after a retro-rocket landing which employs no parachutes. Its inaugural mission will carry cargo only. *(SpaceX)*

ABOVE Montage showing an eroding mesa of sedimentary rocks in Gale Crater, Mars, with an astronaut figure to the correct scale. Mount Sharp rises behind. Much of the planet has the appearance of a desert on Earth. *(NASA/Seán Doran)*

RIGHT The surface of an active comet, with its extremely low gravity and flying particles, would be a difficult place for astronauts to visit. This photo of comet 67P Churyumov-Gerasimenko was taken by the unmanned Rosetta spacecraft. *(ESA)*

envisage a giant spaceship refuelled in Earth orbit taking scores of settlers to the Red Planet.

The two small moons of Mars, Phobos and Deimos, have also been suggested as useful staging posts for missions, and locations from which to study the planet close up without the technical complexity and danger of landing there.

Other planets

Venus, Earth's sister planet in terms of size, is ruled out as a destination for humans by the runaway greenhouse effect of its thick atmosphere which results in surface temperatures of 467°C (872°F) and a pressure 93 times that of the Earth. Mercury has an airless cratered landscape much like the Moon, but its proximity to the Sun and slow rotation means one side is baked in extreme heat and radiation, while the other is plunged into freezing darkness. Further out beyond Mars, the planets Jupiter, Saturn, Uranus and Neptune are all gas giants with no solid surfaces that we know of.

In theory any solid body such as a larger asteroid or planetary moon could be landed on and visited by astronauts in the future. In fact, SpaceX claims its Dragon 2 capsule is designed to be capable of landing anywhere in the solar system because it uses retro-rockets rather than parachutes, and so may descend on to bodies whether or not they possess an atmosphere.

There are many moons in the outer solar system with solid surfaces, often covered in highly useful water ice, which astronauts could potentially explore in the distant future. Jupiter has four major moons, including Europa that has a shell of ice over an ocean. But the Jovian system is a lethal radiation environment meaning that only Callisto, the most distant from Jupiter of the large moons, would be safe for a human visit. Despite being further away from us, the moons of Saturn, Uranus and Neptune are more attractive in this regard.

ABOVE Callisto, the outermost of Jupiter's four Galilean satellites, is the only one whose radiation environment would be tolerable to future astronaut visitors. It is about 40 per cent wider in diameter than the Earth's Moon. *(Montage of Voyager 1 images, NASA)*

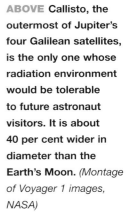

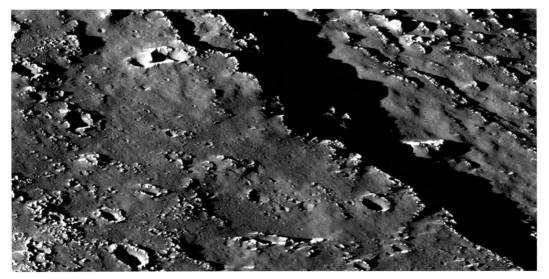

LEFT The rough surface of Callisto is composed of dirty ice, and was photographed up close by the Galileo space probe in 1996. A giant fault or scarp splits the surface, and the largest visible craters are about 0.6 miles (1km) across. *(NASA/JPL)*

Chapter Three

Selection and recruitment

The early cosmonauts and astronauts of the USSR and USA were selected from the ranks of military jet pilots and went on to become national heroes, a path also followed more recently by China. Later, the recruitment pool was greatly widened and strict physical requirements relaxed. Now, the USA, Russia, the European Space Agency, Canada and Japan recruit astronauts openly from the public, although high standards naturally still apply.

OPPOSITE The seven Mercury astronauts pose in front of a McDonnell Douglas F-4 Phantom, an echo of their professional backgrounds as military test pilots. Left to right; Scott Carpenter, Gordon Cooper, John Glenn, Gus Grissom, Wally Schirra, Alan Shepard and Deke Slayton. *(NASA)*

The first people selected to be cosmonauts and astronauts were male military jet pilots, accustomed to flying dangerous missions at high speed, often in perilous conditions and in some cases in combat. This occupation was finally settled as the logical source of candidates in excellent physical condition, with good eyesight, fast reactions, and a degree of fearlessness without recklessness. However, this was only after a number of less obvious occupations had also been considered, including deep-sea divers, racing car drivers and, bizarrely, bull fighters.

Being accustomed to experiencing and handling fear was certainly a necessity, as recruits were to enter a poorly understood and dangerous environment of extreme vacuum, radiation, meteoroid bombardment and other unknown perils. Some exceptional individuals, undoubtedly the minority, actually said that they had never experienced fear – not that they did not know what danger was, and tried to avoid its consequences, and worked hard to get out of hazardous situations, but they did not actually feel frightened.

As space travel became a little more frequent, albeit certainly not routine, engineers and medical doctors joined the ranks, when one of each profession flew on three-man craft Voskhod 1 in 1964. Engineer Konstantin Feoktistov and medical doctor Boris Yegorov joined commander pilot Vladimir Komarov for a one-day flight to a new orbital altitude record of 209 miles (336km).

Valentina Tereshkova became the first woman in space in 1963 by flying the final Vostok mission. However, her apparent illness in orbit,

ABOVE The American space agency NASA and the Canadian Space Agency openly advertise for astronauts in periodic marketing campaigns. *(NASA/CSA)*

BELOW The USSR's Voskhod 1 mission in 1964 was the first to carry cosmonauts who were not military pilots.

BELOW Parachutist Valentina Tereshkova was belatedly recruited to the previously all-male Soviet Vostok program to demonstrate that women could cope with spaceflight.

which was initially covered up, seemed to set back the cause of women in space, rather than encourage an investigation into how to cope with some adverse effects of being weightless. As a result, her successor, Svetlana Savitskaya, did not fly until 1982. The first American woman in orbit, Sally Ride, flew on board a Space Shuttle a year later. Since then, women have flown regularly is space, the vast majority being American. However, in the overall statistics of spaceflight, females remain a distinct minority, comprising 11 per cent of all space travellers.

In addition to the historic space pioneering countries of Russia and the USA, today China, Europe (through ESA, the European Space Agency), Canada and Japan recruit astronauts. All except China are involved in training for the ISS and various ferry vehicles, currently the Russian Soyuz, formerly the US Space Shuttle, and soon a variety of new spacecraft. Each nation's agency has lengthy and thorough selection processes to identify healthy and psychologically stable individuals. Some of the private spacecraft operators are also building small teams of astronaut employees, primarily to pilot their spacecraft.

Astronaut and cosmonaut candidates require the right personality to work well in a team while confined in a small space for extended periods of time, where interpersonal tensions can readily arise and privacy can be limited.

ABOVE This rare photo shows the first woman cosmonaut at the launch pad on 15 June 1963, with three of her fellow recruits. The evening before her flight they were paraded in front of the Vostok 6 rocket and the launch crew. Right to left: Valentina Tereshkova, her back-up Irina Solovyeva, Valentina Ponomaryeva, and Zhanna Yorkina partly obscured behind Ponomaryeva. Chief designer Sergei Korolyev in white shirt is mainly hidden behind Tereshkova. *(Photogallery Leninsk-Baikonur)*

BELOW Sally Ride (1951–2012) became the first American woman in space on Shuttle *Challenger* mission STS-7 in 1983. She flew again on *Challenger* a year later on STS-41-G. *(NASA)*

RIGHT Uniform patch
of the European
Astronaut Centre,
Cologne, Germany.
The centre is the base
for ESA's astronaut
corps. (Spaceboosters)

ABOVE The US Federal Aviation Administration
awards 'Commercial Space Transportation'
astronaut wings to private sector pilots who have
travelled in space. Like the NASA and military
badges, at its centre it has the astronaut device
of a shooting star passing through a halo. (US
Department of Transportation)

RIGHT Gold lapel pin
presented by NASA
to US astronauts
after their first flight.
It is worn on civilian
clothing, and they
have to pay for it. After
completion of training,
they receive a silver
pin of identical design.

even if they never get a flight. This is because
NASA considers the term astronaut to be a job
description for individuals selected to join the
Astronaut Corps at the Johnson Space Center
in Houston, Texas. Once an astronaut candidate
completes training, he or she receives a silver
astronaut lapel pin and becomes a career
astronaut, irrespective of having flown in space.

However, on returning from their first spaceflight
they receive a superior distinctive astronaut badge
and gold version of the lapel pin. There are various
versions of the badges, including the newest one
issued by the US Federal Aviation Administration
(FAA) to the new breed of commercial astronauts,
but they all feature a shooting star flying though an
elliptical orbit symbol.

Many of the early astronauts truly were
extraordinary individuals. A comparatively small
number of people around the world achieve the
top level of their chosen field – be it academic
research, music, medicine, engineering, art,
business, law or whatever specialism they have
entered. But not many at the pinnacle of their
profession would also be highly accomplished
in another unrelated area – say, a top
businesswoman who was also a cello virtuoso, or
a computer scientist who excelled at team sports.

Many of the first astronauts and cosmonauts
did, in fact, fall into this super-elite category,
being accomplished engineers, physicists or
artists as well as experienced pilots in fast jets,
which was the initial route to becoming an
astronaut. Extraordinarily, some actually excelled
in three disparate fields.

Moonwalker Buzz Aldrin was a jet pilot with a
PhD in the arcane subject of orbital rendezvous,
and was an accomplished pole vaulter who

Once selected, they go through a long and
complex training program, which varies according
to the role intended. Some of the topics are
of a general nature, some are specific to the
operation of the spacecraft they will fly in, and
other elements relate to the specific mission role
of the individual in question – for instance as a
commander, pilot, cargo handler, repair worker,
doctor, astronomer or some other specialism.

Many astronaut and cosmonaut candidates
have discovered that their aspirations for a
spaceflight are not guaranteed. Some of those
selected will wait several years to fly in space and
an unlucky few may never make it. The training
program is long and they may fail courses or suffer
a medical issue. Then a flight for which someone
has been training for years can be abruptly
cancelled in the wake of a space accident, a
budget cut, or a change in program plans.

In the USA at least, those recruited and in
training are still regarded as astronauts even
before they make their first spaceflight; indeed,

could still vault at the age of 38. As well as being a test pilot working on experimental aircraft, his crewmate on Apollo 11, Michael Collins, was a top handball player, using this as a final workout before his lunar voyage. Alexei Leonov, the first spacewalker, and American moonwalker Alan Bean, who began as military jet pilots, are both accomplished and published artists.

The first Soviet cosmonauts

The first Russian cosmonaut team was recruited in 1959 from the ranks of pilots serving in the Soviet Air Force. At Luostari Pechenga air base, near Murmansk, on the Kola peninsula, some 900 miles (1,500km) north of Moscow, excited rumours spread as 12 pilots were interviewed, among them Yuri Gagarin and Georgi Shonin. Shonin was asked how he would feel about "flying something of a completely new type".

Six were selected for further medical tests in Moscow, among a large group from around the USSR, which was soon whittled down to 20 cosmonaut candidates. In early 1960, they convened at an old military base north of Moscow, re-named the Cosmonaut Training Centre, later to become the famous Star City (*Zvezdny Gorodok*). The group soon became known by chief Soviet designer and program director Sergei Korolyev as his *Little Eagles*.

They had to be no more than 5ft 7in (170cm) tall, be less than 30 years of age, weigh around 11st 3lb (157lb, 71kg), and be completely healthy with no chronic disorders and even without surgical scars. They were to be fighter pilots with a high-level of flight training, but no specification of flying hours – in fact, Gagarin had 230 flying hours, Gherman Titov who would become the second human to orbit the Earth had 240 hours, and Alexei Leonov, later the first spacewalker, had 250 hours. Finally, as required by the politics of that era, they had to be a member of the Communist Party of the Soviet Union and be ideologically acceptable.

Soviet secrecy meant the full list of individuals and their fates was unknown for over two decades, but in the end, 12 of the original 20 would fly in space – five aboard Vostok spacecraft, three in the Vostok-derived

LEFT Yuri Gagarin was a 26-year-old fighter pilot based at Luostari Pechenga air base near Murmansk in the Soviet Arctic, when he was recruited as a cosmonaut.

LEFT Georgi Shonin, a pilot colleague of Gagarin, was asked by his recruiters how he would feel about flying something 'completely new'. He had to wait six years after Gagarin's orbital trip for his own spaceflight on Soyuz 6.
(*Spacefacts.de*)

LEFT Sign at the entrance to Russia's space centre of Star City (*Zvezdny Gorodok*) reads simply 'Starry'. It houses the Gagarin Cosmonaut Training Centre.

ABOVE Russia's first cosmonauts at Sochi on the Black Sea in May 1961. *Back row, from left:* Filatyev, Anikeyev and Belyayev. *Middle row:* Leonov, Nilolayev, Rafikov, Zaikin, Volynov, Titov, Nelyubov, Bykovsky and Shonin. Front row: Popovich, Gorbatko, Khrunov, Gagarin, chief designer Sergei Korolyev, Korolyev's daughter and child, Karpov (training head), Nikitin (parachute training) and medical chief Yevgeny Fedorov. *(Colin Burgess)*

THE FIRST RUSSIAN COSMONAUTS AND THE PROGRAMS THEY FLEW ON

	Vostok	Voskhod	Soyuz	Salyut	ASTP
Yuri Gagarin	■				
Gherman Titov	■				
Andrian Nikolayev	■		■		
Pavel Popovich	■		■		
Valery Bykovsky	■		■		
Vladimir Komarov		■	■		
Pavel Belyayev		■			
Alexei Leonov		■			■
Viktor Gorbatko			■	■	
Yevgeny Khrunov			■		
Georgi Shonin			■		
Boris Volynov			■	■	

and short-lived Voskhod program, and four having to wait for their first flight until the Soyuz program. All 12 became legendary pioneers of the Russian space program.

■ Yuri A. Gagarin
■ Gherman S. Titov
■ Andrian G. Nikolayev
■ Pavel R. Popovich
■ Valery F. Bykovsky
■ Vladimir M. Komarov
■ Pavel I. Belyayev
■ Alexei A. Leonov
■ Viktor V. Gorbatko
■ Yevgeny V. Khrunov
■ Georgi S. Shonin
■ Boris V. Volynov.

Their space careers worked out as shown in the chart, and included, in addition to the world's first cosmonaut, flying such distinctive missions as the second human orbital flight (Titov), the first joint flight of two spacecraft (Nikolayev and Popovich), commanding the first multi-crew spacecraft (Voskhod 1), and the first spacewalk (Leonov).

The Vostok cosmonauts from this first group

were Gagarin, Titov, Nikolayev, Popovich and Bykovsky. To these five would later be added Valentina Tereshkova, the first woman in space.

The three Voskhod cosmonauts were Komarov, who commanded the first mission with two civilians, and spacewalker Leonov and his commander Belyayev. Komarov later lost his life when he flew the ill-fated first Soyuz mission. Bykovsky had the unusual experience of resuming his space career 13 years after his Vostok 5 mission by making two Soyuz flights, one of which was to a Salyut space station. Leonov went on to command the Russian side of the Apollo-Soyuz Test Project.

The other eight members of the original group failed to make it into space through a combination of training accidents, misbehaviour and subsequent dismissal.

After Gagarin's pioneering flight in 1961, chief designer Sergei Korolyev began the recruitment of female cosmonauts to demonstrate that women could also travel successfully in space. An initial 400 candidates were reduced to five, using criteria including that they be less than 30 years of age, under 5ft 7in (170 cm) tall, the same as their male counterparts, and weigh less than 11st (154lb, 70kg). Furthermore, they had to be experienced parachutists. Valentina Tereshkova was selected in February 1962, and flew on Vostok 6 the following year.

CCCP AND THE USSR

Students of spaceflight history will often see the curious letters CCCP on old photographs of Russian space equipment. The four letters were emblazoned across the front of the space helmet of Yuri Gagarin and his Vostok successors, and can also be seen on the helmet of Alexei Leonov as he made the world's first spacewalk in 1965. They appeared on the shoulder patches of cosmonauts visiting the Salyut and Mir space stations, and on the hulls of Soviet spacecraft.

In fact, the C here does not represent the third letter of the English alphabet, but is the 19th letter of the Cyrillic alphabet and corresponds to the Latin S. The P likewise is the 18th Cyrillic letter, corresponding to R. Hence CCCP in Russian equates to SSSR in English and other languages which use the Latin alphabet including French, German and Italian.

CCCP is an abbreviation formed from the initial letters of the name of the Soviet state which governed Russia and satellite territories up to 1991 – Союз Советских Социалистических Республик.

In the Latin alphabet, this is Soyuz Sovetskikh Sotsialisticheskikh Respublik, or SSSR. Fully translated into English, it is the Union of Soviet Socialist Republics, or the USSR, often referred to in shorthand as the Soviet Union.

So that is the linguistic origin of the CCCP logo commonly seen in historic Russian space photographs, an acronym which should properly be pronounced as 'SSSR'.

LEFT Alexei Leonov making the first spacewalk in March 1965. *(Roskosmos)*

BELOW Commander Vladimir Dzhanibekov (front) and Flight Engineer Viktor Savinykh launched in Soyuz T-13 on a daring mission to rescue the disabled Salyut 7 space station in 1985. Dzhanibekov displays a CCCP flag patch on his left arm.

America's astronauts

The American approach to selection was somewhat more public. In early 1959 NASA secretly set out to select astronauts for its Mercury project. President Dwight Eisenhower ordered that they should recruit military test pilots because such men were already far ahead of other possible candidates in terms of their technical skills, their medical records were already on file, and they were familiar with security requirements.

Qualifications sought included a university degree in science or engineering, and graduation from a test pilot school. Other attributes required were to be no taller than 5ft 11in (180cm), in excellent physical and mental condition and under 40 years of age. Being military pilots who flew new and modified aircraft to evaluate their performance, such men were already used to dealing with risk in a profession which, in the 1950s, was killing its members at a rate of about one per week. To qualify for consideration as an astronaut, a pilot was expected to have 1,500 hours of jet-flying experience.

Comparing the early Russian and American criteria, it is clear that the Americans could be up to ten years older, 4in (10cm) taller, and have extensive jet-flying hours; in fact six times the hours that many of the early cosmonauts had clocked up. The university degree issue was not a constraint on the Russian candidates, but it had unfortunate consequences in eliminating the best US test pilot, the man who broke the sound barrier in 1947, Chuck Yeager.

The NASA selection process continued with screening the records of US Air Force, US Navy and US Marine Corps pilots. An initial 500 possibilities were reduced to 110 candidates. Thirty-five of them were summoned to appear in civilian clothing at the Pentagon in Washington DC. A similar number made the trip a week later. Some failed the height test or other examinations, while others were offered a place and turned it down because the crazy idea of riding rockets into space would be a distraction from their flying careers. The selectors were left with the names of 32 highly qualified individuals – 15 from the Air Force, 15 from the Navy, and two from the Marine Corps. With more than sufficient volunteers, it was decided not to interview a third group. After further tests, the number was whittled down to the final group, which became known as the *Mercury Seven*.

Their names were announced at a crowded press conference on 9 April 1959 in Washington's Dolley Madison House, the post-presidency home of James Madison and now the temporary headquarters of the recently formed National Aeronautics and Space Administration (NASA). The astronauts, who had not even met each other properly before, found themselves on the receiving end of a raucous bombardment of applause, shouted questions and camera flash-bulbs. It was their first taste of the extraordinary publicity, adulation and hero-worship that would soon accompany them wherever they went – before they had even done anything! Lined up behind a table, a few nervously smoked cigarettes as they bemusedly surveyed the baying crowd of reporters and photographers in front of them.

Their names would become iconic in America's Mercury, Gemini and Apollo programs:

- M. Scott Carpenter
- L. Gordon Cooper Jr
- John H. Glenn Jr
- Virgil I. ('Gus') Grissom
- Walter M. Schirra
- Alan B. Shepard Jr
- Donald K. ('Deke') Slayton.

BELOW The 'Mercury Seven' American astronauts in front of a Mercury space capsule. *From the left:* Gordon Cooper, Walter Schirra (behind), Alan Shepard, Virgil Grissom, John Glenn, Donald Slayton and Scott Carpenter.

The table shows how their space careers worked out, and the group included the first American in space (Shepard on a sub-orbital flight), and the first American in orbit, Glenn. Although three would make it through to the Apollo program, Grissom died on the launch pad in the Apollo 1 fire so did not fly on Apollo. Shepard was the only one of the original Mercury Seven to walk on the Moon. Having been grounded by medical issues for 13 years, Slayton finally flew on the Apollo-Soyuz Test Project and met up in orbit with one of his original Soviet counterparts, Leonov. At the age of 77, Glenn made a second flight, on the Space Shuttle, 36 years after his Mercury mission.

Not all of those excluded from the selection of the original Mercury Seven gave up on an astronaut vocation. Jim Lovell and Charles 'Pete' Conrad would succeed in the second NASA intake and go on to stellar careers as astronauts. Ed Givens was accepted much later, in the fifth group, but died in a car crash before he was assigned a spaceflight.

Three years later, in 1962, NASA recruited a second group of nine astronauts, this time by an open application process. The height criterion was relaxed by an inch (2.5cm) as the newer US spacecraft would be roomier than the Mercury capsule. As well as Lovell and Conrad, the successful candidates included many names who would become famous as mainstays of the Gemini and Apollo programs – Neil Armstrong, Frank Borman, Jim McDivitt, Elliott See, Tom Stafford, Ed White and John Young. See was killed in an air crash while training to command a Gemini mission. Seven became commanders of Apollo missions, three of whom walked on the Moon, and one, Young, even went on to command Shuttle flights in the 1980s.

Even these would not be enough to fill the spaces on NASA's ambitious Moon program, so a third intake of 14 astronauts followed in October 1963. This time, being a test pilot was not an essential requirement although jet fighter experience was. Four died while training, but the remaining ten all flew on Apollo, five of them gaining earlier experience in the Gemini program. Four would join the elite band of moonwalkers.

THE FIRST AMERICAN ASTRONAUTS AND THE PROGRAMS THEY FLEW ON

	Mercury	Gemini	Apollo	ASTP	Shuttle
Scott Carpenter	■				
Gordon Cooper	■	■			
John Glenn	■				■
Virgil Grissom	■	■	■		
Walter Schirra	■	■	■		
Alan Shepard	■		■		
Donald Slayton				■	

LEFT Edward Givens was one of several American astronauts to die while they worked towards a spaceflight. Recruited with the fifth intake in 1966, he died in an automobile accident in Texas in 1967 while serving on the Apollo 7 support crew. *(NASA)*

BELOW NASA's second group of astronaut recruits poses with models of Apollo and Mercury spacecraft. *Back row, from left:* Elliot See, Jim McDivitt, Jim Lovell, Edward White and Tom Stafford. *Front row:* Charles 'Pete' Conrad, Frank Borman, Neil Armstrong and John Young. See was killed in an air crash preparing for a Gemini flight, and Ed White died training for Apollo 1, after becoming the first US spacewalker. The remaining seven went on to become core to the Gemini and Apollo programs. Three would walk on the Moon, and three more would orbit it. *(NASA)*

Since those early days, Russia too has recruited its cosmonauts by a more transparent selection process. Today, NASA's Astronaut Selection Program says: "Many men and women have pursued and realised their dreams of flying in space. They all began by submitting their applications to become astronauts."

Current cosmonaut recruitment in Russia

The first open selection drive for Russian cosmonauts took place in 2012, and sought to recruit five or six candidates. Applicants needed to be no older than 33, have graduated from college and have at least five years of work experience. People with backgrounds in humanities are also welcome to apply, so long as they can master the appropriate space technologies. There are of course rigorous psychological and medical tests and conditions, but not as extreme as for the

first cosmonauts. For instance, surgical scars are now permitted, as long as the patient has fully recovered.

The weight range is 110–209lb (50–95kg). Height is no longer the critical factor it once was, and a wide range of statures is now accepted, between 4ft 11in (150cm) and 6ft 3in (191cm). However, travelling in the Soyuz capsule requires an extra dimension of the human frame to be determined – the person's height when sitting reclined in the Soyuz launch chair. Cosmonauts are required to sit with raised knees, feet tucked in and angled necks, and in this posture their so-called *sitting height* is constrained to between 2ft 7in and 3ft 3in (80–99cm). A specific clearance is required, not just to reach the controls in front of the cosmonauts, but because their seats spring upwards upon landing to deploy shock absorbers that soften the impact.

Once the basic criteria are fulfilled, further selection stages include one-on-one psychological checks, physical tests, examinations to measure the candidate's learning abilities and a month-long medical examination. As there is now a range of different professions travelling into space, some have their own special standards – be it for pilots, engineers, researchers or medical doctors. The successful candidates enter the cosmonaut preparation program, which takes at least six

BELOW As American space activity ramped up, NASA recruited a third group of astronauts in 1963. *Back row, from left:* Michael Collins, Walter Cunningham, Donn Eisele, Ted Freeman, Richard Gordon, Russell Schweickart, David Scott and Clifton C. Williams. *Front row:* Edwin Aldrin, Bill Anders, Charles Bassett, Alan Bean, Eugene Cernan and Roger Chaffee. Four would die in training, five flew on Gemini missions, ten flew on Apollo, and of those, four walked on the Moon. *(NASA)*

years, meaning that the 2012 intake would not get a chance to fly into space before 2018.

Former cosmonaut Sergei Krikalev, currently head of the Yuri Gagarin Cosmonaut Training Centre, has pointed out that about two-thirds of applicants come from within the existing space industry, which is traditionally the main source of cosmonauts. The list also includes applicants from other professions, including airline pilots and army officers, plus several women.

The 2012 recruitment drive did not, however, attract anything like as many applications as recent American ones have. The job of cosmonaut is demanding, exciting and rewarding, but the salary scale for even a top-grade cosmonaut is far from lucrative.

Current US astronaut recruitment

NASA has a well-publicised Astronaut Selection Program that opens from time to time on an as-needs basis to select highly qualified individuals for human space programs. It only recruits US citizens, but accepts foreigners who later became American citizens.

Both civilian and military personnel are eligible for consideration. Applicants must meet certain minimum requirements, which include:

■ Bachelor's degree from an accredited institution in engineering, biological science, physical science, or mathematics
■ The degree must be followed by at least three years of related, progressively responsible, professional experience or at least 1,000 hours of pilot-in-command time in jet aircraft
■ An advanced (post-graduate) degree is desirable and may be substituted for experience. A Master's degree counts as one year of experience and a Doctoral degree counts as three years. Teaching experience is considered relevant, and educators are encouraged to apply.

There are notable limitations on the types of Bachelor's degree acceptable to NASA for astronaut applications. While the following fields are related to engineering and the sciences, they are *not* suitable: degrees in

technology (engineering technology, aviation technology, medical technology); degrees in psychology (except for clinical psychology, physiological psychology, or experimental psychology); degrees in nursing; degrees in exercise physiology or similar fields; degrees in social sciences (geography, anthropology, archaeology); and degrees in aviation and aviation management.

All candidate astronauts must meet additional physical requirements, including the ability to pass the NASA long-duration spaceflight physical examination, which includes the following:

■ Height between 5ft 2in (157cm) and 6ft 3in (191cm)
■ Distant and near visual acuity correctable to 20/20 in each eye (with spectacles)
■ Blood pressure not to exceed 140/90 when seated.

By comparison with the poor response in Russia, a NASA recruitment drive in 2012 received over 6,000 applicants, of whom 1,000 were female. This was greatly exceeded in

BELOW Former cosmonaut Sergei Krikalev, executive director of piloted spaceflight at Russian space corporation Roskosmos, explains that two-thirds of Russia's cosmonaut recruits come from within the existing space industry, but public recruitment has now also started. *(Roskosmos)*

2016 when more than 18,000 people applied to be astronauts, beating the record of 8,000 set in 1978 as the Space Shuttle program approached its first launch. It is thought that internet outreach raised public awareness of the posts and more general interest in NASA, and contemporary releases of blockbuster astronaut movies is likely to have helped.

After the preliminary screening of the submissions received, both the civilian and military applicants who are under final consideration go through a week-long process of personal interviews, medical tests, and orientation. An 18-month assessment process then follows, during which NASA chooses between eight and 14 candidate astronauts. The final selected group then go forward for two years of initial training on spacecraft systems, spacewalking skills, teamwork, Russian language, and other requisite skills.

With limited capacity available for newlyrecruited astronauts to visit the ISS, it is not clear what future activities they can expect to undertake, even when new US spacecraft become operational. They will certainly work on systems design and evaluation, and will hope to be assigned to missions to the Moon, asteroids and Mars.

China

China, the third nation to launch its own astronauts, would seem to have followed the same initial recruitment approach as its predecessors.

As the USSR and USA made huge strides in spaceflight through the 1960s, China, which had already joined the others as a nuclear power in 1964, did not wish to be left behind. It commenced its first human space program in 1967 with the aim of putting a two-man crew into orbit in a Gemini lookalike craft called Shuguang.

Nineteen astronauts had been selected by 1971 from among PLA air force pilots. Candidates were required to be 5ft 3in

RIGHT Yang Liwei, China's first astronaut, was selected for flight duties in 1998 and launched into orbit on Shenzhou 5 in 2003.

RIGHT The Charter of the European Astronaut Corps sets out ESA's Vision, Mission and Values for human spaceflight. *(ESA)*

(1.59m) to 5ft 8in (1.74m) tall, weigh 121–154lb (55–70kg) and be between 24 and 38 years old. They required at least 300 hours of flight time – more than the Russian pilots, but much less than the Americans. Detailed screening was based on flying abilities, psychological, physiological, ideological and general medical factors.

The re-entry and recovery of trial uncrewed spacecraft from orbit was accomplished, making China the third nation to achieve this difficult feat. The Shuguang manned spacecraft was planned to be launched on the Long March 2A rocket in 1973, but the program was cancelled in 1972 after losing political support. The program left Chinese space efforts with a useful legacy in the Space Flight Medical Research Centre and communications and tracking ships for recovery of returning space vehicles.

Activity resumed in earnest in 1994 when China bought and licensed space technology from Russia, including the Soyuz design, docking mechanisms, life supports systems, spacesuits and training capacity. Two Chinese astronauts went to train at the Yuri Gagarin Cosmonaut Training Centre in Star City, near Moscow, then returned to China to commence training their colleagues. Twelve astronauts were selected for flight duties in 1998, with Yang Liwei becoming China's first astronaut in 2003. Seven of his colleagues have also flown in space since.

Another group of seven newly qualified astronauts was announced in 2010. It included two women, Liu Yang and Wang Yaping, who flew space missions in 2012 and 2013 respectively. At the time of their selection, all of this group were between 30 and 35 years of age and married. All were pilots in the People's Liberation Army Air Force (PLAAF). The men were jet pilots, the women pilots of transport aircraft. Qualification standards have risen since the early days, as all now have a Bachelor's degree and their average flight time logged is 1,270 hours.

European Space Agency

In 1998, the European Space Agency (ESA) made the decision to form one single European Astronaut Corps out of the separate national efforts in human spaceflight. The first parts of the International Space Station (ISS) were launched that year, and European plans for participation meant a new era was beginning. The European Astronaut Charter acknowledges diversity of nationalities, skills and backgrounds in pursuing the 'common goal of peaceful human space exploration for the benefit of humankind at large and for the European people in particular'.

ESA has only undertaken three astronaut selection campaigns since the agency was created in 1975. The first recruitment drive was in the late 1970s, the second in 1992, and the third ran for a nine month period over 2008–09 that brought in six new ESA astronauts – Timothy Peake of the UK, Alexander Gerst of Germany, Andreas Mogensen of Denmark, Thomas Pesquet of France, and both Samantha Cristoforetti and Luca Parmitano of Italy.

The European Astronaut Corps is based at the European Astronaut Centre in Cologne, Germany. As of 2016, it has 16 members. They may be assigned to work on space projects in Europe (such as the European Space Research and Technology Centre, ESTEC) or abroad at NASA's Johnson Space Center or Russia's Star City. Europe does not have independent access to space, and relies on other nations with spaceflight capability for transportation, currently Russia, and in the past and the likely future, the USA.

ESA does not predict in advance when its next recruitment drive will take place. However, Frank Danesy, Head of Human Resources at the European Space Operations Centre (ESOC) in Darmstadt, Germany, offers some advice: "What I can say is that of those that actually became astronauts, they prepared perpetually, as though the next recruitment campaign was just around the corner."

United Kingdom

These days, any prospective UK astronaut candidates will, like the second Briton in space, Tim Peake, go through the ESA recruitment program.

The case of the first British astronaut was rather different. Helen Sharman was working as a chemist for the Mars chocolate company when she heard an advert on the radio with a simple message of a startling opportunity: 'Astronaut wanted: no experience necessary.' Of the 13,000 applicants, four were sent to Moscow for the final evaluation. Clive Smith and Royal Navy Surgeon Gordon Brooks were

BELOW The 2009 intake of European astronauts comprised *(from left)*: Luca Parmitano (Italy), Alexander Gerst (Germany), Andreas Mogensen (Denmark), Samantha Cristoforetti (Italy), Tim Peake (UK) and Thomas Pesquet (France). *(ESA)*

eliminated. Sharman was selected to fly the Project Juno mission in 1991 with a Russian space crew, with Army Major Timothy Mace as her back-up. Britain thus became the first nation whose inaugural astronaut was a woman, a distinction since repeated by Iranian and South Korean female astronauts.

Canada

Canada has considerable space experience, with the Canadian Space Agency (CSA) recruiting and training astronauts for space missions with the US and Russia. In fact, Marc Garneau became the first Canadian in space in 1984 aboard an American Space Shuttle, five years before the CSA was established. He has since been followed by ten others. One, Robert

LEFT Dr Helen Sharman OBE, formerly a chemist with the Mars confectionery business, became the first British astronaut in 1991 after responding to a promotional advertisement. It would be almost 25 years before she was followed by the first 'official' UK astronaut, Tim Peake.

BELOW The crew of Soyuz TM-12 before launch to the Mir space station in May 1991. From left, Commander Anatolii Artsebarskii, Helen Sharman and Sergei Krikalev.

Thirsk, spent six months at the ISS in 2009, and Chris Hadfield was commander of the ISS for a similar period in 2013, achieving global recognition for singing David Bowie's song *Space Oddity* in the weightless environment of orbit, and for writing a best-selling book.

The last recruitment drive in 2016 took over 3,700 applications, about 70 per cent of whom were male. Ontario, Quebec and Alberta produced the most applications. The CSA's requirements are very similar to those of NASA in that it seeks exceptional people with excellent health, a university education in science, engineering or medicine, and extensive relevant knowledge and experience. The final selection of two astronauts is expected in late 2017. Canada's astronauts mainly work at NASA's Johnson Space Center in Houston, Texas, but also undertake related activities in their home country because Canada supplied the robotic manipulator, called Canadarm2, that operates on the exterior of the ISS.

Japan

With 12 astronauts having flown in space, Japan is in the top league of human spaceflight nations, along with Canada, China and Ukraine (albeit most flew as citizens of the USSR). Japanese astronauts are required to have expert knowledge of science and technology, and to be fluent in English. Selected procedures centre on the screening of application documents, written exams in the English language, interviews, assessing general and specialised knowledge of the natural sciences, and medical and psychological examination.

The first Japanese citizen in space was the journalist Toyohiro Akiyama, who flew on a commercially funded trip to Russia's Mir space station in 1990. The nation's first official astronaut, Mamoru Mohri, flew on Shuttle mission STS-47 *Endeavour* in 1992. Like other early Japanese astronauts, he was selected and trained by the National Space Development Agency of Japan (NASDA), whose functions have since been absorbed into the Japan Aerospace Exploration Agency (JAXA). JAXA astronaut Koichi Wakata became the first Japanese commander of the ISS in 2014.

MISSION PATCHES

The spacesuits, flight suits, and overalls (or coveralls) of astronauts and cosmonauts are usually notable for the colourful mission badges or *patches*, often several, which adorn them. A space mission patch is an embroidered cloth reproduction of a mission emblem, and the tradition seems to have its origins with the military pilots who became the first space flyers. The creation and wearing of patches has now spread to ground crew, recovery teams and contractors.

The first space mission patch was worn by cosmonaut Valentina Tereshkova for her Vostok 6 flight in 1963. Her design features a white dove holding an olive branch of peace. It was on the left shoulder of her thermal garment, under the orange pressure suit worn by all the early cosmonauts, so is not evident in the photographs of her on launch day.

As she entered space she radioed to Earth "Ya Chaika, ya Chaika!" announcing her call sign, "I am Seagull!" Years later she would claim that the white bird on her embroidered patch was a seagull. However, it seems clear from its short beak gripping the traditional olive branch that the design, made by two women who worked on spacesuits, was intended as a dove.

In 1965, the Gemini IV crew were the first American astronauts to wear the Stars and Stripes flag as a shoulder patch. Then Gemini V became the first American mission to have its own patch, to supplement the name badges and NASA logos that the astronauts had on their spacesuits. Gordon Cooper came up with the image of a covered Conestoga wagon of the American West to symbolise their attempt at an eight-day spaceflight record, and it was worn by him and Pete Conrad on the right chest, below their name tags.

Since then, some retrospective patch designs have emerged for the earlier Vostok, Mercury and Gemini flights, but these were not original and do not appear in contemporary photographs of the cosmonauts.

A special style of badge was used for the USSR's Interkosmos program, which took cosmonauts from several nationalities into space; it featured the flag of the participating nation alongside that of the USSR. Towards the end of the Gemini program, crew names were usually added to the patch alongside the symbols designed to represent the mission.

In Apollo, all the flight patches had crew names, except two. For Apollo 11, as the historic first Moon landing, the crew decided not to add their names, intending the design to represent everyone who had worked on the project. Their names and signatures did, however, appear on a plaque they left on the Moon. On Apollo 13, instead of crew names there was the mission's Latin motto *Ex Luna, Scientia* (From the Moon, Knowledge).

Nowadays, all Russian, American, Chinese and European crewed space missions have their own mission patches with crew names shown. Since the fire that claimed the lives of the Apollo 1 crew during training, American patches for use on flight suits have been made from fireproof Beta cloth.

Two women technicians created an artistic patch on the shoulder of Valentina Tereshkova's flight suit. It was hidden under the pressure garment, so was not widely seen. This replica space patch was recreated by a Dutch artist and SpacePatches. nl. It depicts a white dove clutching an olive branch, flying over a backdrop of golden sun-rays. Valentina always claimed it was a seagull, from her call sign 'Chaika', but an ornithologist might disagree. *(Luc van den Abeelen)*

Apollo 15 was the third lunar landing, and its crew chose a design showing their landing site by the Apennine Mountains, three wings to represent the all-USAF crew, and their mission number spelled out in Roman numerals as XV in crater shadows. Worden's is the white wing, flying in orbit above the pair who landed. The initial design came from Italian fashion designer Emilio Pucci. *(NASA)*

In 1978 Czechoslovakia became the third nation to have a citizen in space when Vladimir Remek flew with Soviet cosmonaut Aleksei Gubarev on Soyuz 28 to the Salyut 6 space station. The Interkosmos mission patch has the letters of the USSR and ČSSR (Czechoslovak Socialist Republic) in Cyrillic script.

Space Shuttle mission STS-133 in 2011 was the third to last one of the program, and the last flight of *Discovery*. Its patch was the final work of space artist Robert McCall, completed after he died by Tim Gagnon and Jorge Cartes. Six white stars represent the crew members. *(NASA)*

Shenzhou 5 was China's first human spaceflight, taking Yang Liwei into Earth orbit for 21 hours in 2003. *(CNSA)*

The patch for the Principia mission of British astronaut Tim Peake was chosen via a schoolchildren's competition. It has an apple, referring to Isaac Newton's discovery of gravity, a Soyuz rocket and a stylised space station glinting on the shiny fruit. *(ESA)*

Chapter Four

Training for space

<hr />

Astronauts train in groups as well as individually, to cultivate the teamwork that is an essential part of operating in the dangerous environment of space. They need to master spacecraft systems, water survival for emergency escape, spacewalking, handling remote manipulator devices and, for the International Space Station, foreign languages.

OPPOSITE In the US Gemini program, it was discovered that an underwater environment can provide a useful simulation of weightlessness. Like all modern astronauts, Shuttle and Space Station crew member Peggy Whitson trains in the Neutral Buoyancy Laboratory in Houston, Texas. *(NASA)*

Following the selection of astronaut candidates, the chosen group commences their training. Being an astronaut requires careful preparation, in order to cope with the variety of conditions and dangers which will be encountered, and to enable them to achieve their mission objectives safely. This involves years of training, some of it general for all astronauts, and some of it specific to the role or mission in question.

General astronaut training is almost always done in groups whose members will usually have been recruited in the same intake, and because teamwork is an important part of functioning as a fully fledged astronaut.

Training starts with a general grounding in basic astronaut skills, after which specific training will be required, tailored to a particular specialist technical role or a specific spaceflight mission, once the astronaut is allocated to a flight. Because so many astronauts of

different nations fly on the same spacecraft to the International Space Station these days, primarily the Russian Soyuz, their training has many similarities whether it us undertaken with NASA, the European Space Agency (ESA) or Roskosmos in Russia. Training specific to the Soyuz ride up to and down from the ISS is undertaken at Star City (*Zvyozdny Gorodok*) in Russia. Since the 1960s, Star City has been home to the Yuri Gagarin Cosmonaut Training Centre (GCTC).

NASA Astronaut Candidate training program

In NASA, for example, selected applicants are assigned to the Astronaut Office at the Johnson Space Center (JSC) in Houston, Texas, with the designation of Astronaut Candidates.

BELOW Japanese astronaut Kimiya Yui (rear), Soyuz commander Oleg Kononenko (middle) and Kjell Lindgren (USA) training in the Soyuz simulator for their TMA-17M flight to the ISS in 2015. The training manual cover reads 'Launch and Descent'. The crew took part in ISS Expeditions 44 and 45. *(JAXA/GCTC)*

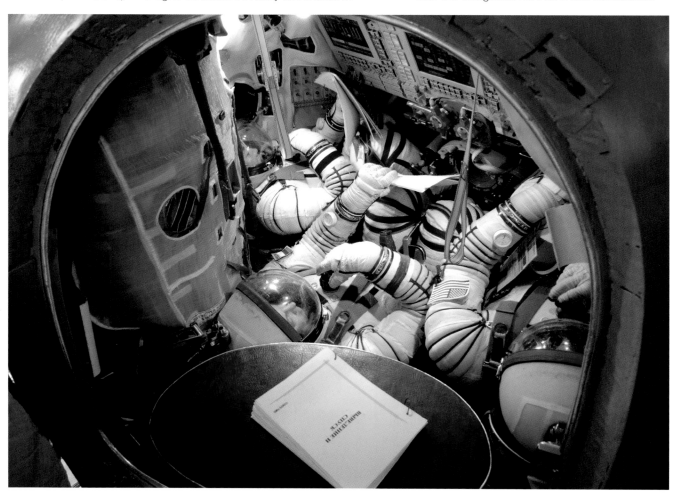

Here, they undergo a training and evaluation period lasting approximately two years. The program is designed to develop the knowledge and skills necessary for formal mission training after they are assigned to a specific flight. Military Astronaut Candidates with jet pilot experience who will actually fly a spacecraft have special program content to enable them to maintain flying proficiency in NASA aircraft during their training period.

Candidates have to study training and operations manuals, and undertake computer-based training lessons on the various spacecraft systems. Specialist training staff take them through the systems in detail, explaining how to operate them, and how to recognise and rectify malfunctions.

Training in water is an important part of the program. Many spacecraft splash down in the sea or occasionally lakes, either planned like the US Mercury, Gemini and Apollo capsules, or as an unplanned contingency, like Soyuz and Shenzhou. The new SpaceX Crew Dragon capsule will also initially splash down in the ocean, before propulsive ground landings are attempted later. Space capsules carry life rafts and other safety flotation equipment.

Hence Astronaut Candidates are required to complete military water survival training. Since underwater diving is somewhat similar to going outside the spacecraft in a spacesuit, they become qualified as SCUBA (self-contained underwater breathing apparatus) divers during their preparation for spacewalk training. The candidates have to pass some specific swimming tests during their first month of training. This involves swimming three lengths of a 25m pool without stopping, and then swimming a further three pool lengths in a flight suit and tennis shoes, without a time limit. They also need to demonstrate that they can tread water continuously for ten minutes wearing a flight suit.

Water training in a giant tank in the Neutral Buoyancy Laboratory at JSC is also used to simulate spacewalking. The trainee wears a full spacesuit with its flotation finely balanced, so it neither sinks nor rises, but floats as if the person inside were actually weightless. It is not a totally analogous sensation since, unlike in the vacuum of space, there still remains the effort required

ABOVE NASA's Lyndon B. Johnson Space Center (JSC) in Houston, Texas, formerly the Manned Spaceflight Center (MSFC), is where US human spaceflight training, research, and mission control are undertaken. The call-sign 'Houston' was frequently heard throughout American spaceflights.

LEFT Checklists are a constant part of astronaut training and operations. This one covered the launch phase of Space Shuttle *Atlantis* on STS-135, the final mission of the Shuttle program in 2011. *(NASA)*

LEFT Canadian astronaut Julie Payette, mission specialist on STS-127, undergoes water survival training in the Neutral Buoyancy Laboratory. She is wearing a training version of her orange Shuttle launch and entry suit. *(NASA)*

CENTRE In the US Gemini program, it was discovered that an underwater environment can provide a useful simulation of weightlessness. Here, Gemini XII astronaut Edwin 'Buzz' Aldrin rehearses his spacewalk at the Agena docking target vehicle. *(NASA)*

to move limbs and body against the resistance of the water. However, the astronaut acquires some familiarity with the dynamics of their own body motion under weightlessness, and is able to rehearse EVA activities planned for orbit.

Candidates are given experience of exposure to the dangers associated with high (hyperbaric) and low (hypobaric) atmospheric pressures in altitude chambers, where they learn to deal with the emergencies that can arise in these conditions.

Full-scale mock-ups of segments of the ISS are used for more realistic familiarisation and training. Here astronauts can rehearse how to operate experiments and cameras, and undertake more mundane but essential tasks such as preparing food, dealing with waste and stowing equipment. Unless such matters are handled in an organised way, weightless items can readily float away and get lost or cause problems with sensitive equipment.

RIGHT The Russian Hydrolab water immersion facility at the Gagarin Cosmonaut Training Centre in Star City, near Moscow. The tank contains mock-ups of the Russian segment of the ISS, where cosmonauts train for spacewalks. *(Christopher Michel/ Creative Commons)*

Being on a sub-orbital flight in the company's SpaceShipTwo (SS2) vehicle, the astronauts will only have a few minutes of weightlessness. Hence, an important part of the preparation is to become familiar with how to enjoy free fall, and maximise their viewing of the Earth through the SS2's 12 windows while traversing the apex of their flight.

They also must take appropriate care to be comfortably and safely positioned in their personalised recliner seats during the higher g-forces experienced in the descent phase. Although they will float weightless out of their seats around the cabin when in space, they need proper cushioning as the spacecraft rapidly decelerates upon re-entry to the atmosphere.

The Virgin Galactic operation employs aerospace medical staff to ensure each astronaut has been through the required medical check-ups and is safe to fly. Travelling to space does of course carry some inherent risk, on account of the speeds and energy levels required to make the ascent.

Training for the Moon

The surface of the Earth's Moon constitutes a very different situation for the astronaut from Earth orbit, lunar orbit or any other familiar space environment, whether this be inside a spacecraft such as the ISS or venturing outside on a spacewalk or EVA. However, it also has a number of important similarities to orbital operations, and some of the astronaut's skills and experience can be usefully transferred.

In the 1960s NASA, the only space agency to have achieved a crewed lunar landing to date, used a variety of simulation techniques to familiarise the astronauts assigned to lunar landing missions. Following this training, a total of 17 astronauts eventually flew in the nine Apollo Lunar Modules (LM) which carried crews

ABOVE Chinese crew member Wang Yaping takes a cardiovascular endurance test in preparation for the Shenzhou 10 mission to the Tiangong-1 space station in 2013. *(Qin Xian'an)*

RIGHT NASA's Lunar Landing Training Vehicle (LLTV) was used by the commanders of the Apollo Moon flights to practise the final stage of the descent. Here Charles 'Pete' Conrad rehearses his upcoming Apollo 12 landing at Ellington Air Force Base, Texas. *(NASA)*

in space. Six of those LMs actually landed on the Moon, two were used in tests in Earth and lunar orbit, and in the case of Apollo 13 the crew were forced to abandon their intended landing mission. Eugene Cernan, the last man on the Moon, was the only individual to fly in two LMs.

Accurate replicas of the forward part of the LM cabin were built at NASA's Manned Spacecraft Center (now the Johnson Space Center) in Houston, Texas. A terrain model of the landing site was made from plaster of Paris, and a small television camera on a gantry moved over the landscape in response to inputs to the LM controls by the astronauts. The images created were projected on to the windows of the simulator to give the impression of descending towards the Moon.

Actual flying training for the Moon was undertaken in helicopters because, although the Moon is airless, a low-velocity, near-vertical landing in a rotary wing aircraft provides a good analogue for a landing in a vacuum when the LM is powered by a rocket firing vertically downwards. Cosmonauts on Russia's cancelled Moon program, including Alexei Leonov, also trained for a lunar landing on a modified version of the Mi-8 helicopter.

More practical training using jet power was done on the Lunar Landing Training Vehicle (LLTV). Made by Bell Aerosystems, this ingenious craft had a large jet engine pointing downwards to cancel five-sixths of its weight, and so simulate lunar gravity. The pilot used smaller hydrogen peroxide rockets to control the descent and mimic a lunar landing. Test pilot Joe Walker, who had earlier become an astronaut in his own right (albeit only posthumously recognised) in the X-15 rocket plane, undertook the operational trials for the craft.

The LLTV was somewhat unstable and unreliable, but nevertheless was highly regarded as a training tool by the Apollo commanders. Three of these vehicles were destroyed in crashes, but fortunately the pilots were able to eject to safety. One of those was Neil Armstrong, who used it extensively to train for the first lunar landing, and despite his mishap said it was an excellent simulator which greatly contributed to the success of his mission.

Gravity

A key point to note when devising astronaut operations on the Moon is the nature of that body's gravitational field. Being smaller than the Earth, with a diameter only one quarter (27

BELOW John Young, commander of Apollo 16, gives a 'big Navy salute' as crewmate Charlie Duke photographs him at their Descartes landing site in the lunar highlands. As the shadows reveal, Young has leapt a couple of feet (about half a metre) off the ground from a standing start in the low lunar gravity. In the background are the Lunar Module *Orion*, the lunar rover and Stone Mountain, about 4 miles (6km) away. *(Charles Duke/NASA)*

per cent) of the parent planet, and a mass of only 1.23 per cent, the gravitational force on the lunar surface is 16.5 per cent of Earth – meaning objects there weigh almost exactly one-sixth of what they would on Earth.

This may at first seem surprisingly strong, bearing in mind that gravity depends on the mass of the attracting body, and the mass of the Moon is so small compared with the Earth. It is also much less dense, at 60 per cent of the Earth's mean density. Like the Earth, the densest part sits in the core. The other factor in determining the surface gravity is that the force also depends on distance from the centre of the body, and objects on larger planets are much further away from the core than on smaller ones.

For an astronaut standing on the lunar equator, the entire mass of the Moon lies no more than 2,160 miles (3,476km) away from him, exerting the full gravitational attraction of the body. For someone on the Earth's equator, that distance would not even reach the core, far less the other side. The Earth's densest gravitation attractor, the metallic core, has its centre 3,963 miles (6,378km) below the surface, and the rocks on the far side of the planet are 7,926 miles (12,756km) away, exerting a much feebler pull. Hence, the Moon's surface gravity is greater than might be expected from consideration of its small mass alone.

This has all sorts of implications for astronaut activities, and the design of lunar spacecraft, equipment and instruments. Support structures, ladders, cables and so on do not need to be designed to the same strength as they would be for use on Earth. However, because most of them will be manufactured on Earth, at least until 3D printing becomes more widespread, they will need to retain their rigidity under 1*g*. They also must survive the additional acceleration experienced during launch and landing, so the savings in material required will be less than if an object can be locally manufactured on the Moon.

Atmosphere and dust

Because of its low gravity, the Moon is unable to hold on to much of an atmosphere, which escapes off into space or is more directly stripped away by the solar wind, the stream of energetic particles coming off the Sun. The

result of Apollo experiments left on the Moon showed the Earth's satellite to have a total atmosphere weighing only about 10 tonnes. When a lunar lander descended to the surface, the exhaust gases from its rocket engine doubled the total quantity of lunar atmosphere! Hence operations on the lunar surface are effectively conducted in the vacuum of space. This means there is also intense and potentially deadly radiation bathing the lunar landscape, unfiltered by any atmosphere.

What is more, meteoroids and micrometeoroids rain down sporadically on the surface, with no gaseous atmosphere to burn them up by friction. They travel in an unimpeded path at high speed to the surface, creating little craters called "zap pits" in the soil and rocks, or if an astronaut is unlucky, hitting their helmet, tearing their suit or damaging equipment.

Lack of a blanketing atmosphere to transfer warmth by convection also affects lunar temperatures. In direct sunlight, the Moon's surface can reach a temperature of over 123°C (253°F). At night and in daytime shadows, the temperature can drop to minus 180°C (minus 292°F), and considerably lower in permanently shaded craters around the

ABOVE Among the hazards of moonwalking is the danger of impact from micrometeoroids. Small ones leave microscopic pits like these on Alan Bean's helmet, which were revealed by taking impressions back on Earth. In the main image, Bean holds an environmental sample container filled with lunar soil, while Apollo 12 crewmate Pete Conrad is reflected in his visor. *(NASA/ Charles Conrad)*

poles – an instrument on a recent NASA orbiter determined the temperature of the surface in some polar craters as minus 238°C (35 degrees above absolute zero, or minus 396°F).

Bombarded by meteoroids for millennia, the lunar surface has been pulverised into a dry mixture of shattered rocks, angular gravel and extremely fine dust. This usually covers the bedrock with a layer of soil, or regolith, several metres deep. At every step, the astronaut raises a volume of dust. This does not form swirling clouds, as it would on Earth, because that requires an atmosphere. Instead, each little particle travels off on its own individual ballistic trajectory, bouncing off its neighbours, on a sweeping arc to land some distance away. The slightest movement by an astronaut sends this material flying upwards and outwards in all directions.

This makes working out on the lunar surface a messy affair, and astronauts on the later Apollo missions, which spent three days on the

Moon, soon found their spacesuit zippers to be clogged with dust. When back inside the cabin of their landing craft, the fine lunar dust got everywhere, even on to their faces. It gave off the distinct smell of cordite.

Suited for the Moon

The implications of the lunar environment for the astronaut are that they need the same degree of protection in a pressure suit as they would for a spacewalk in Earth orbit, and in some ways more. Their garment will need to cope with high temperatures in sunlight, and the polar opposite in shadow. It requires a tinted helmet visor to protect the eyes against unfiltered sunlight, and a thick outer layer to their suit to shield against micrometeoroids. In addition to the normal weightless EVA gear, they will also need detachable over-boots with a deep tread to grip the surface and provide the traction required for safe mobility in an environment where astronauts have their full mass, but only one-sixth of their weight.

Whereas the oxygen supply used by Alexei Leonov during his pioneering spacewalk, and later by the EVAs of the American Gemini missions and the first American space station, Skylab, was supplied by an umbilical that connected their suits to the vehicle, to have used such a scheme for walking on the Moon would have imposed an impossible constraint on movement. Hence, the backpack approach to carrying consumables, as currently used on the ISS, is essential for lunar EVAs.

Adding a backpack full of breathing oxygen, cooling water, pumps, pipes and radio equipment to an already heavy spacesuit results in a considerable weight. The entire outfit used by the 12 Apollo moonwalkers in the 1960s and '70s weighed 180lb (82kg) before it was strung around with a camera, tools and rock sample bags. That is approximately equal to the weight of the astronaut inside it. In lunar gravity, of course an astronaut weighs one-sixth of his Earth weight, and enjoys this pleasant sensation inside the cabin while stripped down to his underwear or when sleeping in a hammock. But out on the surface, the suit brings the total weight up to one-third of his Earth weight, and also greatly restricts his movement.

In fact the Apollo suit is light compared with the US Shuttle suit and its life support system, which at about 310lb (140kg) is 70 per cent heavier. However, this is where mass, reflecting the amount of effort required to move bulk, comes into play irrespective of low or zero gravity. In the weightless environment of orbit this does not matter, except when the spacewalker needs to float along and make sufficient effort to shift the total mass of himself, suit and pack. But the moonwalker standing upright has the full lunar weight of the suit on his shoulders, and has to start, control, and stop the momentum of double his own mass as he moves around. In the low gravity, one might not realise one has leant over too far. Hence the danger of toppling over is always present, as happened to several moonwalkers, fortunately without ill effects.

Working on the Moon involves a range of tasks related to equipment, field surveys, photography and pulling or driving wheeled vehicles. Experiment packages need to be carried to an appropriate site, deployed with the aid of spirit levels, and connected to a central station which will collect and transmit data to Earth. Rock and soil samples have to be collected using rakes, tongs or scoops, and chipped from rock faces with a geology

BELOW Lunar over-boots were worn during the 14 moonwalking excursions undertaken by the Apollo astronauts. These were the last to tread the lunar surface, used by Eugene Cernan, commander of Apollo 17. *(Smithsonian Air and Space Museum)*

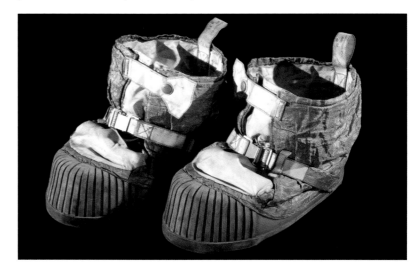

ABOVE The traction obtained in lunar gravity can be simulated by a trainee being suspended, then walking along a surface inclined at 9.6° off the vertical. This is the Reduced Gravity Simulator at the NASA Langley Research Center. *(NASA)*

BELOW Three months prior to the Apollo 11 mission, Neil Armstrong practises soil sampling for the short time he would spend on the lunar surface. *(NASA)*

hammer. A hammer proved to be a versatile tool. As Alan Bean advised during a moonwalk, "Don't come to the Moon without a hammer!"

The aspiring lunar astronaut trains for these tasks in a combination of one-sixth gravity simulators and in normal Earth gravity. The early simulators used counterweights to offset five-sixths of the trainee's weight, giving the feeling of only weighing one-sixth of normal. Another method had the trainee suspended sideways in order to walk on an inclined board set 9.6° off vertical to accurately simulate the traction between a boot and the lunar surface.

An altogether more realistic sensation was provided by flying the Vomit Comet aircraft on an arc that provided brief periods of one-sixth gravity rather than the normal weightlessness. This was used to test the first wheeled vehicle used by humans on the Moon, the MET (Modular Equipment Transporter) that was used on Apollo 14, also known as the "lunar wheelbarrow".

Before they abandoned their secretive lunar program, Soviet cosmonauts started training in 1968 in preparation for a Moon landing. A moonwalk simulator was installed in the gymnasium in Star City, outside Moscow.

ESA Shuttle astronaut Jean-François Clervoy performed an underwater simulation of a moonwalk by adjusting his buoyancy to simulate lunar gravity. After trials in the pool of diving company Comex, he made two moonwalk simulations in the Mediterranean Sea off Marseilles, collecting soil samples with tools similar to those used on the Moon by the Apollo crews. This underwater test was a step towards developing European expertise in spacewalk simulations under partial gravity for exploring the Moon, asteroids and Mars.

NASA also has a more modern one-sixth gravity chair which mimics the experience of moonwalking, and is used in Space Camp.

Most lunar training is done in normal Earth gravity. The Apollo astronauts wore training spacesuits and tried out their tools using a simulated dusty lunar surface, peppered with rocks of various sizes. They rehearsed selecting and photographing rocks, lifting and bagging them, raking out pebbles and digging trenches.

As with Earth orbit and lunar orbit, and unlike missions to asteroids or Mars, the modern

Chapter Five

Spacecraft

Modern spacecraft are of two types, capsules and spaceplanes, while the ultimate goal of a single-stage vehicle flying to orbit remains tantalisingly unattainable. With the retirement of the Space Shuttle, America is returning to capsule designs, including commercially designed and operated ones, for its future access to space. The astronaut, whether pilot or ordinary crew member, needs to master a multitude of systems and may well fly on more than one type of spacecraft.

OPPOSITE A visualisation of Boeing's CST-100 Starliner in flight. The Starliner owes much to the design of the Apollo Command Module of the 1960s, but its chief role is to ferry humans to low Earth orbit rather than heading for the Moon. *(NASA/Boeing)*

The aspiring astronaut has to learn the systems and operation of his spacecraft. The extent to which this is required, and the level of detail involved, will depend upon the precise role of the astronaut.

Despite increasing automation, commanders and pilots need to know how to fly their spacecraft in every conceivable situation, overcome a wide range of anomalies and potential emergencies, and understand how the systems operate. Astronauts whose

principal jobs entail payload handling, science investigations or undertaking spacewalking repairs, can get by with less detail, but must be able to operate certain cabin equipment, understand their responsibilities in an emergency and be ready to be assigned specific extra duties by the commander.

The pilot astronaut or systems specialist will have access via radio, video and email to a team of engineers on Earth who will collectively have a much greater level of knowledge of the spacecraft than the astronauts could ever assimilate. However, they cannot always rely on this support. If a situation develops suddenly, or if there is a failure of telecommunications, or in the period of radio blackout during re-entry to the atmosphere, then the crew of a spacecraft are on their own and must deal with any anomaly by themselves.

There are two basic types of human spacecraft in current use, capsules and spaceplanes, and both can be used for either orbital or sub-orbital flight.

Capsules are basically hardened shells without wings, whose shape has been designed to protect their occupants from both the vacuum of space and the searing heat of re-entry. They are launched by a multi-stage rocket and descend by parachute and/or landing rockets. This landing may be by parachute to a controlled splashdown in the sea without rockets, or by a combination of parachutes and rockets on terra firma.

Spaceplanes also require a rocket, either single or multi-stage, to launch them into orbit, but their method of return is quite different from that of a capsule. They have wings, and advanced heat protection surfaces that enable their complex aerodynamic shapes to survive re-entry and then be actively flown to a controlled landing on a runway like a conventional aircraft. They touch down on wheels or skids, and reusability is one of their key advantages.

Capsules have been in use since the start of human spaceflight in 1961, and since the retirement of the US Space Shuttle in 2011 there has been a return to this concept. The first two men to orbit the Earth flew in the Soviet Vostok capsule and came down on land. The first two American spaceflights were sub-orbital arcs in the Mercury capsule, with

BELOW Vostok was the first human space capsule, carrying Yuri Gagarin into orbit in 1961, but it flew in unmanned trials the previous year.

a splashdown in the Atlantic Ocean. Since then, all three space-faring nations have flown a wide range of capsule designs, including the US Gemini and Apollo capsules, the Russian Soyuz and the Chinese Shenzhou. Capsules are especially suited for returning from deep space missions because they do not need to carry the extra weight of wings, control surfaces and wheels.

After America's 30-year run of the Shuttle program, capsules are making a comeback. The private sector SpaceX Dragon capsule has flown many times in its unmanned form and a version capable of carrying a crew is imminent. Other NASA-sponsored capsules are in the works from Boeing (CST-100) and Lockheed Martin (Orion), and both Russia and China are testing larger modern capsules capable of carrying more crew than their existing spacecraft. India has also made progress in designing its own crewed space capsule.

These modern capsules are all conical, with a heat shield covering the wide base. They tend to have steeper sides than the original Apollo capsule which pioneered this design, and is currently the only vehicle to have carried people back to Earth in a high-speed re-entry from lunar distances, which it accomplished nine times. Earlier capsules were spherical (Vostok and Voskhod), truncated cones with a cylindrical nose (Mercury and Gemini), or curving tapered hulls (Soyuz and Shenzhou).

The X-15 rocket-powered aircraft was belatedly recognised as the first spaceplane when one of its pilots received posthumous designation as an American astronaut for flying the craft twice above the Kármán Line, the boundary of space, in 1963. The hypersonic X-15 could reach Mach 6.7, or 4,500mph (7,200kph), which is more than one-quarter of the velocity required to enter Earth orbit. Its air-launched mode was first used for the Bell X-1 rocket plane, in which Chuck Yeager broke the sound barrier in 1947.

Various orbital spaceplanes have been designed and tested since the early 1960s, but only finally flew in 1981 with the first launch of the US Space Shuttle, *Columbia*. This remains the only time a human-rated spacecraft has flown with crew on its maiden flight (because it could not have flown unmanned), and its huge

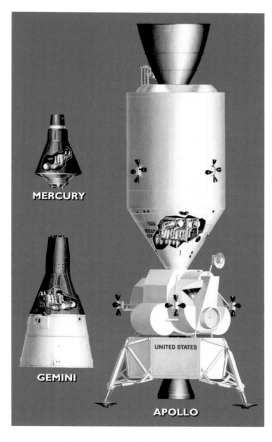

LEFT The three US space capsules from the 1960s carried one-, two- and three-man crews. *(NASA)*

BELOW The X-15 rocket plane was air-launched from a B-52 carrier aircraft. This A-2 model had a white ablative coating to protect against frictional heating, external fuel tanks, and a ramjet test device.

In fact, it is bigger than the first Salyut, Skylab and Mir space stations.

Russia designed a lookalike shuttle of its own called Buran, albeit its propulsion system was quite different from the American original. On its maiden flight without crew in 1989 it was steered back to a successful runway landing. With the collapse of the Soviet Union, the program was abandoned.

Nevertheless, the reusable shuttle vehicle concept remains a viable means of reaching near-Earth space, and can be expected to have an important role to play in the future of spaceflight. Other small reusable shuttle-type vehicles are in design or testing, including the unmanned US X-37B which has spent up to 675 days in orbit at a time, on secret missions for the US Department of Defense. In the private sector, the Sierra Nevada Corporation is developing a spaceplane called Dream Chaser to fly to the ISS, initially unmanned to deliver cargo and ultimately to serve as a crew taxi.

The sub-orbital SpaceShipOne became the first privately owned and funded craft to enter space in 2004, in the process establishing the potential for the new industry of space tourism. Several governments, including that of the UK,

ABOVE The US Space Shuttle was the world's first reusable spacecraft. Here *Discovery* approaches the ISS on mission STS-120 in 2007 with the *Harmony* node in its cargo bay. *(NASA)*

size and cargo capacity compared with earlier spacecraft meant it could carry large loads to and from Earth orbit.

Being roughly comparable in size to a DC-9 airliner and weighing 74 tons (165,000lbs) empty, the Shuttle remains by far the largest human spacecraft ever flown to and from orbit.

RIGHT The X-37B is a small unmanned reusable spaceplane which has spent almost two years in orbit before returning to Earth. *(Boeing)*

are establishing regulations on the certification and safe flight of commercial spaceplanes from their territory.

SSTO

The holy grail of spaceflight, sought by designers since the 1960s and naively thought by many pre-Space Age science fiction writers to be the normal routine in spaceflight, is the concept of Single-Stage-to-Orbit (SSTO). The concept is that a spaceplane or aerodynamic rocket will be fuelled and launched from Earth, travel to orbit, complete a mission such as rendezvous with a space station, and then return home intact. No booster stages are discarded en route, no ablative heat shield is permanently burned off on re-entry and the craft is promptly prepared for another flight after refuelling and a quick check-up. Hence SSTO vehicles hold out the promise of greatly reduced launch costs by eliminating the expensive wastefulness of expendable launch systems, even ones that are partially reusable like the Space Shuttle.

While SSTO has never been accomplished from Earth, calculations suggest it should be marginally possible with advanced propulsion technologies. It is much easier to achieve on smaller planetary bodies with lower gravity and minimal atmosphere, such as Mars and the Moon.

Earth-launched SSTOs might be regarded as a separate, third and special category of spacecraft, as it is so different from both the capsule and the booster-augmented spaceplane. A spaceplane SSTO would probably take off from a runway with a special engine performing first as a conventional jet engine to accelerate beyond Mach 1. Then its shape is transformed into a supersonic combustion ramjet (or scramjet) which compresses and chills incoming air to burn the fuel, rather than carrying its own heavy load of oxidiser, like a rocket. Beyond Mach 5, the final stretch to orbit would need to be performed by a conventional rocket engine.

McDonnell-Douglas produced SSTO designs in the 1960s and revisited the topic in the 1990s with its DC-X rocket. Many others have also worked on SSTO designs, including HOTOL by Rolls-Royce and British

ABOVE SpaceShipOne became the first non-government craft to enter space during a sub-orbital flight in 2004 piloted by Mike Melvill, the first commercial astronaut. *(Ian Kluft)*

BELOW NASA's X-43A experimental unmanned scramjet set the world speed record for a jet-powered aircraft at Mach 9.6 or 7,000mph (11,000kph). It may help revolutionise access to orbit. *(NASA)*

Aerospace and Skylon by Reaction Engines Ltd, and the X-33 VentureStar by Lockheed Martin. Several failed attempts to build working prototypes would appear to demonstrate that the SSTO concept is unattainable using current technology, but is a realistic future prospect.

Current and imminent spacecraft

The retirement of the Space Shuttle, which President George W. Bush decreed would occur once construction of the ISS was complete, posed a dilemma for NASA. The *Columbia* disaster in 2003, coming after the loss of *Challenger* in 1986, sealed the fate of this pioneering vehicle. Although it had demonstrated practical reusability in spacecraft for the first time, it had never achieved the operational economics that should have come with it.

The development of a new Crew Exploration Vehicle (CEV), later named Orion, was well underway when the final Shuttle flight, *Atlantis* on STS-135, landed at the Kennedy Space Center in July 2011. However, the CEV was many years away from its first flight. How would the US get its astronauts to the ISS, whose construction had cost NASA many years of hard toil and billions of dollars?

An interim solution was provided once again by Russia's reliable and repeatedly upgraded Soyuz spacecraft. As in the period 2003–05, when the Shuttle was grounded, American astronauts again started flying to the ISS solely on Soyuz spacecraft. In fact, Russian commanders had been flying American, European and astronauts of other nationalities, both professional and private, to the ISS on Soyuz spacecraft for many years in parallel with Shuttle flights.

This reliance on another nation for human access to space was not a situation which the US found desirable. So NASA turned to the private sector with its Commercial Crew

Program (CCP), setting the goal of achieving safe, reliable and cost-effective access to and from the International Space Station and low Earth orbit. Longer term, NASA aims to lay the foundation for future commercial space transportation capabilities in which private sector businesses will increasingly operate their own routine access to space.

Thus NASA has part-funded two companies, Boeing and SpaceX, as crew transportation contractors under this program, backing them to complete the development of two independent new human-rated space capsules of different designs, and their respective launch systems. Other companies are also involved in different aspects of the CCP, and together with the two main spacecraft designers, this approach is providing NASA with multiple options and flexibility. It reduces overall risk to the program, because if one vehicle is temporarily grounded the other will be able to continue reliable missions to the ISS.

Once the new space transportation system is certified to meet NASA requirements, the agency will fly missions for space station crew rotation and emergency returns. These new spacecraft must be able to carry four astronauts and serve as a *lifeboat* that can safely and quickly evacuate the space station's crew in an emergency. They also must demonstrate that they can serve as a 24-hour safe haven during an emergency in space and be able to stay docked to the station for at least 210 days.

The spacecraft also need a launch escape system (LES) to quickly whisk the crew away from the launch vehicle in the event of a malfunction. This is similar to an ejection seat for a jet pilot, but instead of ejecting the crew from the spacecraft, the entire spacecraft is carried away from the launch vehicle by powerful rockets that are either installed in a tower above the spacecraft or as part of the spacecraft's structure. Once clear, the capsule descends by parachute and/or braking rockets to an emergency landing.

SpaceX and Boeing will use hangars and launch facilities at NASA's Kennedy Space Center and the adjacent Cape Canaveral Air Force Station to prepare their spacecraft, rockets and crews for flight.

Once the companies complete their crewed flight tests, NASA will certify each to launch astronauts on regular crew rotation missions to the space station. With this certification, the contracts call for each company to receive orders for up to six orbital flights by crews of four astronauts. Assuming a successful start, subsequent contracts can be expected to follow.

In time, NASA plans to routinely buy this service, like getting a taxi ride to low orbit. Because the companies will own and operate the systems, they will be able to sell human space transportation services to other customers and thereby reduce their operating costs. The target date for routine commercial crewed flights to the ISS is no earlier than 2018. NASA installed a new docking port on the ISS for use by future commercial craft in 2016, on top of an old Shuttle docking port that is no longer required. In fact, this new port was itself delivered to the ISS by a commercial SpaceX flight.

By encouraging private companies to provide crew transportation services to a region of space that NASA has been visiting since the Mercury mission of John Glenn in 1962, the agency believes it can maximise the research and operational experience it gets from its investment in the International Space Station.

The Boeing Starliner and SpaceX Crew Dragon carry four crew members on each mission, permitting the crew complement on the ISS to increase from six to seven crew members. As a result, the total crew research time on the orbiting laboratory can be doubled from 40 hours per week to 80 hours, thereby greatly expanding science investigations that increase our understanding of what it takes to live and work in space.

Astronauts who will fly commercial missions to the station will work closely with contractor-led test teams throughout the final stages of development and certification of these new spacecraft. This Joint Test Team approach will include a NASA astronaut on the crew of the test flight to the ISS. It continues NASA's longstanding approach of having its astronauts closely involved in the engineering development of their spacecraft. Once certified, NASA will use these services to fly American and US-sponsored astronauts to the ISS.

ABOVE Soyuz TMA-7 transported the crew of Expedition 12, Valery Tokarev and Bill McArthur, along with space tourist Greg Olsen, to the ISS in 2005. Periscope, docking probe and Kurs rendezvous antenna are clearly seen. *(NASA)*

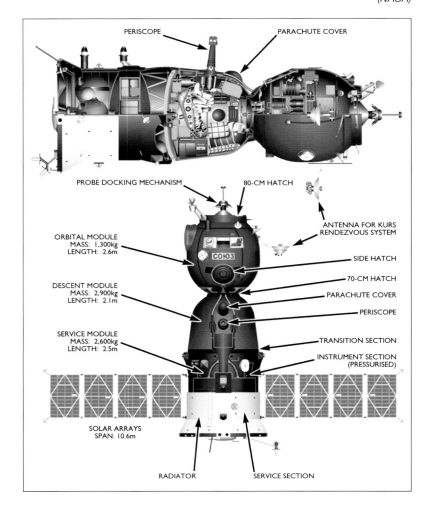

PERISCOPE

PARACHUTE COVER

PROBE DOCKING MECHANISM

80-CM HATCH

ANTENNA FOR KURS RENDEZVOUS SYSTEM

ORBITAL MODULE
MASS: 1,300kg
LENGTH: 2.6m

СОЮЗ

SIDE HATCH

70-CM HATCH

PARACHUTE COVER

PERISCOPE

DESCENT MODULE
MASS: 2,900kg
LENGTH: 2.1m

TRANSITION SECTION

INSTRUMENT SECTION
(PRESSURISED)

SERVICE MODULE
MASS: 2,600kg
LENGTH: 2.5m

SOLAR ARRAYS
SPAN: 10.6m

RADIATOR

SERVICE SECTION

NASA's program to hand over low Earth orbit astronaut transportation to commercial companies is designed to allow the agency to free up resources to develop its architecture for space exploration much further from Earth. In parallel with CPP, NASA also wants to focus on building spacecraft and rockets for deep space missions, including to Mars in the 2030s. Consequently, in parallel with the CCP involving private industry, NASA is continuing to develop its own human spacecraft. Derived from the cancelled CEV, the Orion is intended ultimately for government-funded flight beyond Earth orbit.

Significantly, all three of these post-Shuttle American space vehicle programs – Orion, the Boeing Starliner, and SpaceX's Crew Dragon – have reverted to a capsule approach not unlike the Apollo Command Module whose design dates from the early 1960s.

Capsules

Soyuz MS

The Soyuz MS is the latest and final upgrade planned for the venerable and reliable Soyuz spacecraft, which has been carrying human passengers into Earth orbit since 1967. Its maiden flight was in July 2016, when commander Anatoli Ivanishin flew to the ISS with flight engineers Takuya Onishi of Japan and Kate Rubins of the USA.

Like all earlier Soyuz craft, the MS has three standard modules. An aerodynamic capsule, or Descent Module, accommodates the crew for launch and re-entry, and this is attached to a cylindrical Service Module which houses engines and other support systems and carries two wing-like solar panels. On the front of the Descent Module is a spheroidal Orbital Module, which the crew can enter when in space, and which houses rendezvous and docking apparatus and the hatch through which to enter the ISS.

LEFT Component parts of Russia's reliable Soyuz spacecraft. It has been the backbone of Soviet and Russian human spaceflight since 1967, and frequently the only means of access to the International Space Station. *(ESA/David Woods)*

The MS has a number of modifications over the Soyuz TMA-M which preceded it. There are several new antennae – some for rendezvous and navigation on the Orbital Module, omnidirectional ones at the tips of the two improved solar panels, and the new EKTS communications system carried on top of the Service Module that faces skyward to permit either direct communications with ground stations, or links to mission control in Korolev, outside Moscow, via Russian Luch communications relay satellites. A new GPS and Glonass satellite navigation system (ASN-K) will help to determine the precise position of the spacecraft in orbit, and also enable quicker location of the capsule after landing, the latter being especially useful in the event of an off-course ballistic re-entry.

The new Kurs-NA approach and docking system weighs half as much and consumes one-third of the power of the previous Kurs-A system, and the on-board computer is also new. Higher overall power consumption of all the new devices meant adding a new battery on board.

The location of manoeuvring thrusters on the Service Module has been also altered, and there is a new external lighting system on the Orbital Module. Rather than adopt the speedy ISS rendezvous conducted over four orbits which had become the norm in recent years, Soyuz MS-01 deliberately took a longer trip to permit a two-day checkout of the new design prior to docking. Many of these new systems are likely to be transferred over to Russia's planned replacement for the Soyuz, the PTK-NP which has been named Federation.

Orion MPCV

As well as encouraging the private sector, NASA is continuing to develop its own spacecraft. The Orion is the agency's first new human-rated spacecraft in more than a generation, and is intended to supersede the retired Shuttle in certain roles.

Derived from the CEV of the Constellation program that was started by President George W. Bush in the wake of the *Columbia* disaster, only to be cancelled by his successor, the development of the Orion Multipurpose Crew Vehicle (MPCV) is currently well underway.

ABOVE The control panel in the Soyuz Descent Module has been modernised in recent models of the spacecraft. At top is the hatch to the Orbital Module, with three Kazbek seats below. *(ESA)*

BELOW Orion, NASA's new spacecraft, from the right: Launch escape tower and protective shroud, Command Module, European-built Service Module, four solar panels, three curved panels of the spacecraft adapter to connect to the launch vehicle. *(NASA)*

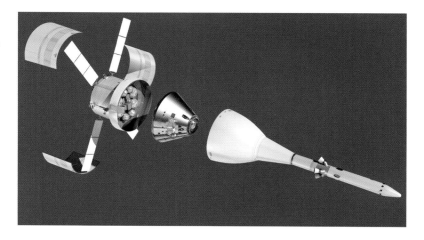

However, because it will be the first human spacecraft designed for deep space missions beyond Earth orbit and the Moon, the testing required to qualify it for such ambitious missions is expected to last until 2023.

The Orion spacecraft has two main modules, a Command Module for crew occupation and a Service Module containing rocket engines, four solar panels, batteries and other essential services. The CM carries a standard crew of four and is built by Lockheed Martin in New Orleans, Louisiana.

The Service Module is of somewhat narrower diameter than the base of the Command Module to which it is attached, and carries the propulsion system and tanks of oxygen and water for the crew. The service module's structure is also designed to provide locations on which to mount scientific experiments and carry cargo. In the new era of international cooperation, NASA is to procure this module through the European Space Agency from Airbus Defence and Space. Its design is based on the propulsion module of ESA's Automated Transfer Vehicle, which made five unmanned cargo flights to the ISS.

During launch, Orion also has a spacecraft adapter consisting of three curved panels to mate the spacecraft to the launch vehicle and to protect the folded solar panels from the supersonic wind stream.

The Orion MPCV borrows basic design elements of the Apollo Command Module of the 1960s and '70s, but is larger, having a base diameter of 16ft (5m) compared with 13ft (3.9m) for Apollo, and offering approximately 320ft^3 (9m^3) of internal capacity compared with 210ft^3 (6m^3) in Apollo. Naturally, Orion's technology and capabilities are much more advanced than the earlier capsule. In particular, it is designed to support long-duration missions in deep space, with up to 21 days' active crew usage, plus six months quiescent, during which crew life support is to be provided by another module such as the planned Deep Space Habitat. The spacecraft's life support, propulsion, thermal protection and avionics are designed to be upgradeable in future, as new technologies become available.

The Orion spacecraft will be launched by the Space Launch System rocket, which is also currently under development, and is intended for missions into deep space and eventually to Mars. NASA regards both destinations – the International Space Station and deep space – as being vital to the success of future space exploration.

On Exploration Test Flight 1 in late 2014, the spacecraft was launched unmanned by a Delta IV Heavy rocket on a 4.5 hour flight with a peak altitude of 3,600 miles (5,800km). Accelerating downwards during its last engine firing, the Orion capsule attained a re-entry speed of 20,000mph (32,000kph; 8,900m/s) in order to expose the heat shield to temperatures up to 4,000°F (2,200°C). It then made a successful splashdown under three parachutes in the Pacific Ocean.

The first crewed Orion flight is not expected until 2023, when the primary objective will be to check out Orion's crew systems close to Earth, especially the life-support equipment.

On the second crewed mission, the craft will fly deeper into space, perhaps passing around the far side of the Moon. NASA's Directorate for Human Exploration and Operations envisions the Moon serving as a prime testing ground for trials, which will be flown at a slow and deliberate pace of one flight per year. A more ambitious potential destination, never previously reached by humans, is a near-Earth asteroid. Docked with a capacious Deep Space Habitat, the Orion spacecraft is to fly to Mars in the 2030s and later to destinations far beyond.

Closer to home, Orion will also be able to

BELOW Astronauts train in the Orion cockpit at JSC, wearing NASA's new MACES spacesuit. *(Robert Markowitz)*

travel to the ISS to supplement the services provided by commercial operators, or to retrieve crew and cargo, if needed. Other LEO destinations, such as attending to satellites, will also be accessible if required.

Sound spacecraft designs can show astonishing longevity, like the Russian Soyuz, which has been carrying crews for half a century, mirroring the durability of successful aircraft like the Boeing 707 and de Havilland Comet, which flew for 60 years. If the Orion proves satisfactory, it could well be flying deep space missions for many decades as an element of a larger deep space transportation system.

SpaceX Crew Dragon

SpaceX made history in 2012 when it became the first commercial company to deliver cargo into orbit to re-supply the International Space Station, and then return cargo to Earth – feats that had previously been achieved only by government space programs. However, the Dragon capsule that accomplished this was also designed from the outset to be capable of being further developed to carry people, and SpaceX is well on the way to accomplishing this.

SpaceX had a head-start over fellow CCP candidate Boeing, in that they had already developed the unmanned Dragon and were using it to make deliveries to the ISS under a separate NASA contract. The new Crew Dragon capsule is to carry four astronauts into orbit and to dock with the ISS. It is rather sleeker-looking, and quite different in external appearance and internal systems from the original cargo carrier.

Crew Dragon is a fully autonomous spacecraft that can be monitored and if necessary controlled by on-board astronauts and the SpaceX mission control in Hawthorne, California. For certain uses, it is capable of launching with up to seven astronauts in total. The crew will wear SpaceX spacesuits for critical manoeuvres as a precaution against unplanned depressurisation.

In preparation for docking, the Crew Dragon hinges its nose cone open to expose the docking unit that is designed to couple with the NASA Docking System (NDS) on the ISS. This mechanism is 63in (1.6m) wide externally, with an internal transfer tunnel diameter of 31in (0.8m) for the passage of crew and cargo. Two

NDS berths, conforming to the international docking system standard, are intended to accommodate the Crew Dragon, Boeing CST-100 and Orion MPCV.

Crew Dragon capsules will be capable of remaining at the ISS for up to 210 days, the same lifetime as the Soyuz, serving as a lifeboat to take the research lab's crew home in an emergency. Following undocking, the nose cone will swing shut to protect the docking apparatus during re-entry.

After a de-orbit burn and jettisoning of the trunk section, the Crew Dragon capsule commences its return to Earth, protected by an ablative heat shield. Inside the capsule is an ingenious moveable ballast sled, which alters the centre of mass in flight and permits more precise control of the trajectory during the descent through the atmosphere. The craft is advertised as being able to land "almost

ABOVE Visualisation of the Orion capsule re-entering the Earth's atmosphere. *(NASA)*

BELOW SpaceX Crew Dragon undergoing engineering work. The capsule flies in orbit with a cylindrical trunk section attached. Pairs of yellow-capped pusher rockets in the black recesses provide launch abort and propulsive landing capability. *(SpaceX)*

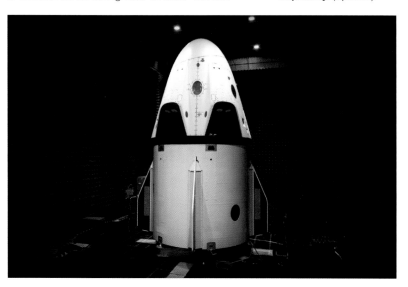

flame-retardant driver seat covering material for Formula One racing cars.

Crew Dragon's displays provide real-time information on the state of the spacecraft and its systems – including its attitude, its position in space, possible destinations and the on-board environment. A tablet-like computer display swivels down for optional crew control. The Environmental Control and Life Support System (ECLSS) provides for adjustable cabin conditions. During a flight, astronauts can set the interior temperature in the range 65–80° Fahrenheit (18–27°C). The spacecraft can return safely to Earth even if there is a cabin leak up to an equivalent orifice of 0.6cm (0.25in) in diameter.

Beneath the Crew Dragon capsule on the launch pad, and attached to it during orbital flight, is the so-called *trunk* section – a cylinder with four fins which supports the spacecraft during its ascent to space, carries unpressurised cargo and houses the solar arrays and radiators for dissipating excess heat. It also provides aerodynamic stability during an emergency abort. The trunk and solar arrays remain attached until shortly before re-entry, when they are jettisoned.

Crew Dragon features an advanced emergency launch escape system (LES) which

anywhere in the world" on four extendable legs with the accuracy of a helicopter. In the event of the primary rocket landing system failing, there is a back-up parachute landing system.

Crew Dragon is designed to be reusable, and SpaceX hopes each capsule will fly ten times prior to requiring extensive refurbishment.

The capsule has five windows providing views for each occupant, unlike on the Space Shuttle. The seats are made from high-grade carbon fibre and Alcantara cloth, a suede-like material which is 68 per cent polyester and 32 per cent polyurethane. This is also used as a

was tested in 2015, to rapidly carry astronauts to safety if something were to go wrong. The crew will experience about the same g-forces as on an adventurous fairground ride.

The integrated pusher-rocket launch escape system is claimed to have several advantages over the tractor motor in a detachable tower used on most previous crewed spacecraft, including Mercury, Apollo, Soyuz and Shenzhou. The stated benefits include the capability for crew escape during the entire flight to orbit, reusability of the escape system, and improved crew safety that results from eliminating a stage separation (meaning jettisoning the tractor unit once its work is done, without which the enclosed capsule could not open its parachutes and descend to safety).

The pusher rocket system also offers the ability to use the same escape engines during descent for a precision Earth landing of the capsule. This rocket system, capable of being used both for an in-flight abort and a nominal Earth landing, employs a cluster of eight of SpaceX's SuperDraco rocket engines. These can hold their fuel for a long duration and be fired more than once. An emergency parachute system will be retained as a redundant back-up for water landing.

After an uncrewed test launch, SpaceX plans to launch its first Crew Dragon with astronauts in 2018, piloted by a company test pilot and a NASA astronaut.

Crew Dragon missions will be launched on Falcon 9 rockets from Pad A of Launch Complex 39 at the Kennedy Space Center. SpaceX leases this historic pad, from which Apollo 11 and most Moon flights departed, from NASA. Subsequently, a version of this spacecraft is to be launched by the more powerful Falcon Heavy on a mission to deliver 2–4 tonnes of cargo to the surface of Mars.

Boeing CST-100 Starliner

Boeing, the other major NASA CCP contractor, is working in parallel with SpaceX to design and test a space capsule called the Crew Space Transportation (CST)-100 that will be capable of carrying seven astronauts to the ISS.

Although Boeing developed the spacecraft for NASA, the company also hopes to carry commercial passengers and to undertake

missions for other national governments. The agreement with NASA permits Boeing to sell seats to space tourists, and the company is working with Bigelow Aerospace and Space Adventures to advance space tourism.

The CST-100's conical Command Module has a diameter of 15ft (4.56m), which is larger than the Apollo capsule but smaller than NASA's Orion capsule. This is attached to a cylindrical Service Module housing engines, electrical power systems, radiators and other equipment. The Starliner capsule has an innovative, weld-less design which Boeing is manufacturing in Florida.

ABOVE Crew Dragon arrives at the ISS, its nose cone flipped open to expose NASA's new docking system. An unmanned Dragon cargo craft is already docked at right. *(SpaceX)*

BELOW Parts of the Boeing CST-100 Structural Test Article in the former Shuttle processing hangar at NASA's Kennedy Space Center, Florida. *(NASA/Kim Shiflett)*

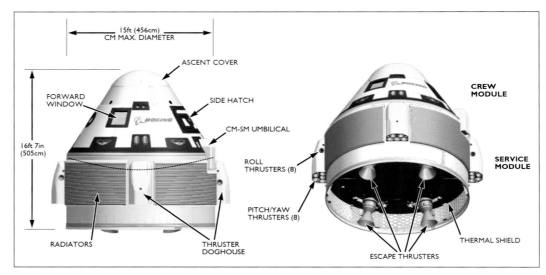

RIGHT The Boeing CST-100 Starliner will keep open NASA's options for access to orbit. (Boeing)

15ft (456cm) CM MAX. DIAMETER

ASCENT COVER

FORWARD WINDOW

SIDE HATCH

CM-SM UMBILICAL

16ft 7in (505cm)

ROLL THRUSTERS (8)

PITCH/YAW THRUSTERS (8)

RADIATORS

THRUSTER DOGHOUSE

CREW MODULE

SERVICE MODULE

THERMAL SHIELD

ESCAPE THRUSTERS

It can accommodate up to seven passengers comfortably, or a mix of crew and cargo. The initial standard NASA requirement is for five passengers plus cargo. In addition to serving the ISS, it is also designed to visit a habitat module which is planned by Bigelow Aerospace.

As autonomous flight and docking will be the norm, this will reduce the training time for crews. The CST-100 has tablet touch-screen technology to enable the crew to monitor the mission and provide input when necessary. The cabin interior has wireless internet, and is lit by LED lighting of the type that is already in use on Boeing 737 and 787 airliners.

The CST-100 has pusher abort engines in the SM, which propel the spacecraft off a malfunctioning rocket during launch. If unused,

as would normally be the case, their fuel is available to provide on-orbit boost to any vehicle the CST-100 docks with, be that the ISS, a Bigelow habitat module or some other vehicle. This is a significant difference from the Crew Dragon, where the abort rockets need to retain their fuel for landing, because the CST-100 abort rockets, being on the SM, will be jettisoned before re-entry.

The CST-100 will use the NASA Docking System at the ISS, and remain attached for up to 210 days to serve as a lifeboat to carry the crew home in an emergency, as well as more routine travel.

At the end of a normal mission, astronauts enter the CST-100 and the craft undocks, moves to a safe distance and performs an automatic de-orbit burn. Protection from the heat of re-entry is provided by the Boeing Lightweight Ablator (BLA). After the period of maximum heating and deceleration, the usual system of drogues and main parachutes will slow the descent. Landing will be on a terrestrial site in the western United States, initially New Mexico or Utah.

For the final stage of the landing, airbags are located underneath the heat shield, which is designed to be separated from the capsule while under parachute descent at about 5,000ft (1,500m) altitude. The airbags are inflated by a mixture of compressed nitrogen and oxygen, to cushion the impact of landing. Although not common, airbag-cushioned landings have been used before, including on the Mercury capsule which deployed a large bag before it hit the

BELOW The Boeing CST-100 capsule will descend under parachutes to land on airbags in the western USA. (BLM Nevada)

ocean. More modern airbag systems were used on the unmanned Mars landings of Pathfinder in 1997 and two Mars Exploration Rovers in 2004. Having been tested in realistic conditions at Delamar Dry Lake, Nevada the CST-100 capsule's airbags are proven technology.

The CST-100 will initially be launched from SLC-41 at Cape Canaveral Air Force Station, Florida, on an Atlas V rocket operated by the United Launch Alliance (ULA). This variant of the rocket will have two solid rocket boosters strapped to either side of the core stage to augment the thrust at lift-off. The upgraded Centaur upper stage, powered by two RL-10 engines, has not flown on the Atlas V before. Also, the crew capsule's launches will be the first time this workhorse launcher has flown without an aerodynamic shroud on its nose. Boeing and ULA engineers conducted wind tunnel tests to try out alternative designs to overcome aero-acoustic issues and verify that the Atlas V and CST-100 are compatible.

However, the Boeing spacecraft is also being made compatible with alternative launch vehicles, including the Delta IV, Falcon 9, and the planned ULA Vulcan. Boeing plans to refurbish the CST-100 Starliner capsules and fly each one up to ten times.

The first full orbital flight in 2018 is to be an uncrewed test mission to the ISS, dubbed Boe-OFT, that will last 30 days. The first crewed flight, Boe-CFT, will follow several months later with a two-person crew consisting of one NASA astronaut and one Boeing test pilot flying a 14-day mission to the ISS.

After commanding the final Space Shuttle flight in 2011, NASA astronaut Chris Ferguson transferred to Boeing to refine the fidelity of a CST-100 simulator by evaluating virtual space manoeuvres. These include manual piloting activities, on-orbit attitude and translation manoeuvres, docking at the ISS, undocking and withdrawal, and a manual re-entry on the way home to Earth. The simulator builds on aircraft training technology and uses actual spacecraft flight software to train Starliner pilots for future missions.

PTK-NP Federation

The convoluted acronym for Russia's new manned spacecraft, PTK-NP, stands for

Pilotiruemyi Transportny Korabl - Novogo Pokoleniya, meaning Piloted Transport Ship - New Generation. Fortunately the Russians have also come up with a concise everyday name, the *Federatsiya* or Federation. It takes this from the Russian Federation, the official name of the country, and echoes the naming origin of the Soyuz spacecraft. That was derived from the Russian word for Union, Soyuz, referring to the nation state at the time, the Soviet Union – but also cleverly hinted at the docking of two Soyuz spacecraft which was planned for later flights of the vehicle.

Whatever we call it, the Federation is built by RSC Energia and comprises a large conical Descent Module atop a narrower diameter cylindrical Service Module, with wing-like solar panels on each side. Its internal volume of 600ft^3 (17m^3) is much greater than Crew Dragon or the CST-100 Starliner, and almost as big as Orion, but the habitable portion of that internal space is bigger than any of them, and can carry up to six crew. A hinged control panel in front of the pilots can be swung out of the way to provide clear access to the forward transfer tunnel that leads towards the docking hatch. Unlike its Soyuz predecessor, there is no Orbital Module.

This spacecraft will be launched on the Angara-5V rocket, a human-rated version of Russia's new rocket which was introduced in unmanned form in 2014. The first launch is planned for 2023 from the new Vostochny Cosmodrome in the far east of the country. In the event of a launch vehicle malfunction, the

ABOVE The Federation, Russia's new space capsule, showing internal pressure vessel, exterior shell and docking probe. *(Roskosmos)*

capsule will be pulled clear by a tractor-type emergency escape rocket on the nose.

Once in orbit, the vehicle will navigate to the ISS using variants of the systems developed for Soyuz, such as the Kurs rendezvous apparatus. Federation has a modified Soyuz docking probe compatible with the Russian ports on the ISS and because there is no Orbital Module to be discarded, the docking mechanism can be returned to Earth with the capsule for re-use.

To further reduce Russia's reliance on Kazakhstan, which hosts the Baikonur Cosmodrome and the primary recovery sites for Soyuz missions, the Federation spacecraft will land at a new site not far from the Vostochny Cosmodrome. Achieving this will require the capsule to manoeuvre during its passage through the atmosphere. After the Service Module has performed the de-orbit burn and been jettisoned, the capsule undergoes re-entry, and then descends by parachute. As the ground approaches, the heat shield is jettisoned to allow four landing legs to deploy from the base of the capsule.

Whereas the Kazbek couches of the Soyuz require an individual body mould to be made for each cosmonaut, the Federation will contain new Cheget shock-absorbent couches that can be adjusted to suit each flyer. Finally, a solid-propellant rocket landing system will produce a gentle vertical landing. The violent Soyuz landing, which some have compared to a "car crash", often followed by the capsule rolling across the ground, will no longer happen.

In the event of a hard landing due to rocket failure or undeployed landing legs, Federation would land on a crushable under-structure that would absorb the impact and be survivable for the crew. However, the capsule, designed to make at least ten flights, would likely be a write-off in such an emergency scenario.

After an uncrewed trial flight, Federation is expected to carry its first cosmonauts to the ISS about 2024. A crewed mission into lunar orbit could follow a year or so later.

Shenzhou

The name of the Chinese spacecraft Shenzhou is not unambiguously translatable into English, but it means something like 'Divine Vessel'.

Superficially, it looks externally similar to Soyuz, and owing to a technology transfer deal with Russia has adopted some elements of that design. This is especially evident in the general layout of its compartments and the specific shape of the Entry Module (equivalent to the Descent Module of Soyuz), which permits some aerodynamic control during re-entry. However, it is larger and made of different materials and construction techniques, and is domestically built.

The fact that the capsule has about 50 per cent more interior volume than Soyuz means that the crew can lie on their backs in a more comfortable arrangement. Chinese software controls the flight on the Long March 2F launch vehicle, and colour television is transmitted throughout the launch phase to enable the

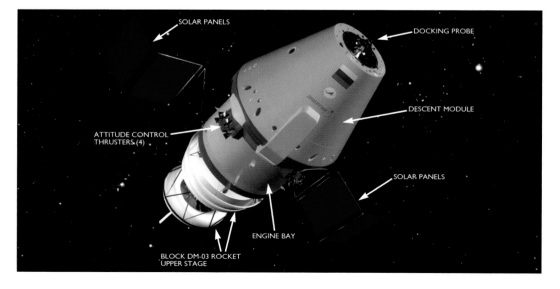

RIGHT Key components of the Federation spacecraft, attached to a Block DM-03 upper stage propulsion unit. *(Roskosmos)*

options and spread the risk by bringing in the cargo version of Dream Chaser to improve its future re-supply capability. This will serve to accelerate plans to make the Dream Chaser an operational option for carrying astronauts.

Other spaceplanes

The Indian Space Research Organisation has flown a winged prototype spaceplane in a hypersonic flight from the Satish Dhawan Space Centre which ended with a controlled, unpowered glide to a simulated runway landing. The reduced-scale test, the Reusable Launch Vehicle Technology Demonstration (RLV-TD) pathfinder, had delta wings and angled tail fins that gave it an outward appearance similar to

the US Air Force's unmanned X-37B orbiter.

The project could eventually lead to future unmanned and manned reusable launch vehicles that could take off like a rocket, ascend to orbit with cargo, return to Earth and land on a runway. An operational reusable launch vehicle based on the winged spaceplane design would have to be at least five times larger than the 4,000lb (1,800kg) test vehicle.

The European Space Agency also tested a spaceplane in 2015, called the Intermediate Experimental Vehicle. This had a mass similar to the Indian vehicle, flew a sub-orbital arc to an altitude of more than 250 miles (400km), and then dived into the atmosphere at a much faster speed than the RLV-TD pathfinder.

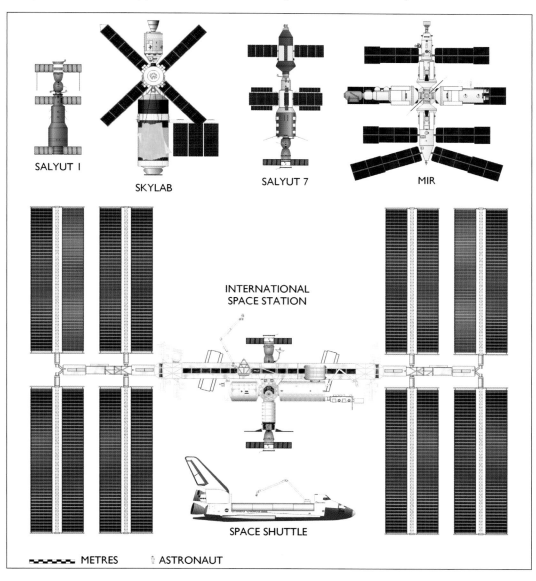

SALYUT 1

SKYLAB

SALYUT 7

MIR

INTERNATIONAL SPACE STATION

SPACE SHUTTLE

METRES   ASTRONAUT

LEFT Today's ISS compared with the US Space Shuttle and four previous generations of space stations – Salyut 1 (USSR), Skylab (USA), Salyut 7 (USSR) and Mir (USSR/Russia).
(Richard Kruse/ HistoricSpacecraft.com)

Chapter Six

Entering space

A launch on Russia's Soyuz rocket is the principal access route to the International Space Station. It is a voyage redolent with traditions that originate from the early pioneers of the Space Age. Astronauts of many nationalities have taken this tumultuous ten-minute ride from the steppes of Kazakhstan to the dazzling environment of Earth orbit.

OPPOSITE Expedition 32 to the International Space Station begins its journey from Kazakhstan on board the Soyuz TMA-05M craft. The Soyuz FG rocket is derived directly from the type that sent Sputnik and Yuri Gagarin on their pioneering flights. *(NASA-Bill Ingalls)*

ABOVE Italian astronaut Paolo Nespoli signs the guest book in Gagarin's office, along with Soyuz TMA-20 crewmates Dmitri Kondratyev and Catherine Coleman, while their back-ups look on.
(ESA-Kargapolov)

There is no easy way to cross the threshold of space and enter the exotic and dangerous realm which has fascinated mankind for centuries. The official boundary is 62 miles (100km) straight up, more than 11 times the height of the world's highest peak, Mount Everest, and far beyond the survivable limit for the unprotected human body.

Getting there safely requires a pressurised cabin capable of preserving its integrity in a vacuum, propelled by, with current technology, some kind of chemical rocket. This is the only route to the cosmos, whether the space traveller follows the gentler trajectory of sub-orbital flight and nudges over the Kármán Line to achieve astronaut status or accelerates all the way up to orbital speed of 17,450mph (28,100kph) at the minimum altitude required to travel all the way around the planet above the atmosphere, which is approximately 100 miles (160km).

Thus far, we have not been able to develop a single-stage-to-orbit vehicle, so everyone who has flown in orbit has left the surface of the Earth in a flurry of flame aboard a multi-stage chemical rocket. This mode of operation is likely to continue for the foreseeable future. Sub-orbital tourists will follow the same and somewhat gentler route as the X-15 and SpaceShipOne, with the spacecraft being ferried high into the troposphere under the wing of an aircraft and then released, firing its rocket to pursue a ballistic arc which peaks just above the boundary of space.

Thus the flights of the new generation of spacecraft now being tested will be achieved by carrying them on launch vehicles of fundamentally the same type as used by the pioneering cosmonauts and astronauts at the beginning of the Space Age.

Although there are several of these new spacecraft on the horizon, and there have been several successful types in the past, at the time of writing there are only two vehicles that can carry humans into orbit, namely Russia's venerable and reliable Soyuz and the Chinese Shenzhou, which owes much of its design to the engineering legacy of Soyuz.

Of the two, Soyuz is by far the more familiar to the modern astronaut. Since the retirement of the American Space Shuttle in 2011, it has been the only way for astronauts of all the nationalities that participate in the International Space Station – Russia, USA, Europe, Canada and Japan – to access it and return to Earth. Hundreds of individuals have already flown safely to space in Soyuz, without a single fatality during launch, although there have been several stressful aborts.

The use of Shenzhou, in contrast, is restricted to Chinese astronauts either for free-flying missions or to visit their experimental Tiangong space stations. Since 2003, Shenzhou has carried about a dozen people into space.

Preparing for Soyuz launch

Although the world's first cosmonaut, Yuri Gagarin, lost his life in an air crash in 1968, his legacy is a prominent part of the modern Russian space program. Crews getting ready to ride the Soyuz launch vehicle, whose first stage is directly derived from that which carried Sputnik and Gagarin, observe a number of traditions which either commemorate the man or repeat his own actions prior to making his historic spaceflight.

The cosmonauts' base of Star City (*Zvyozdniy Gorodok*), which houses the Yuri Gagarin Cosmonaut Training Centre, lies 40km north-east of Moscow. So crews make the short trip in to Red Square to visit the Kremlin Wall and place red carnations in tribute at the graves of the heroes in the Russian space program. Here, on the outside of the wall facing the square, are the ashes of Gagarin, chief designer Sergei Korolyev, cosmonaut Vladimir Komarov who died in the Soyuz 1 crash, the crew of Soyuz 11 and the Salyut space station (Georgi Dobrovolsky, Vladislav Volkov and Viktor Patsayev) and Vladimir Seryogin, the jet pilot who died with Gagarin.

Back in Star City, the crew sign the guest book in Gagarin's office, which is now a museum. The office preserves the contents exactly as they were when he left it for his fatal flight on 27 March 1968.

About two weeks before launch, the prime and back-up Soyuz crews for a mission fly to the Baikonur Cosmodrome in Kazakhstan. In future, they can expect to fly instead to the Vostochny Cosmodrome in Russia's Far East, once that is ready for human spaceflight, but by then they will likely ride into orbit aboard the new Federation spacecraft.

Baikonur is a town and military area formerly known as Tyuratam (*Toretam* in Kazakh) which Russia leases along with a gigantic area of steppe

ABOVE ISS Expedition 49-50 crew members Shane Kimbrough of NASA (left), and Sergey Ryzhikov (centre) and Andrey Borisenko (right) of Roskosmos arrive in Kazakhstan. *(NASA)*

BELOW The town of Baikonur (formerly Leninsk) in Kazakhstan and today's Soyuz launch pad at Gagarin's Start (Ploshchadka No. 1).

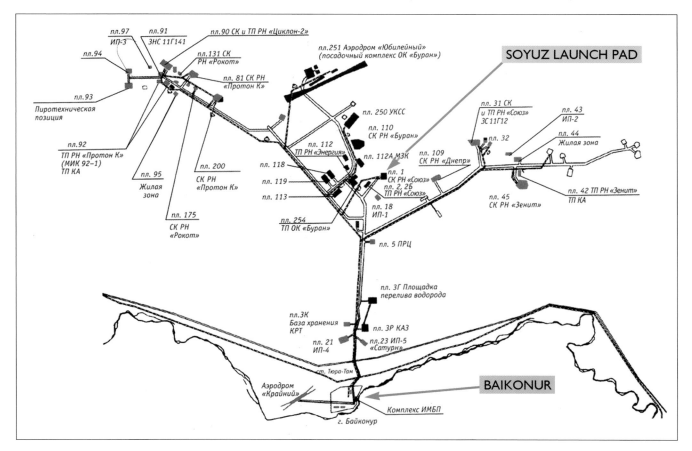

bust of chief designer Sergei Korolyev, statues of Gagarin and Vladimir Lenin, a model Proton rocket, a real Soyuz rocket and other space-related artefacts and monuments.

The crews reside in the Cosmonaut Hotel, which is about 20 minutes' drive from the cosmodrome. In a park nearby is an avenue of trees leading down to the Syr Darya river, and there they perform another traditional ceremony in which each cosmonaut plants a tree. Currently the plot has about 200 trees, each one planted by a person who has launched into space on board a Soviet or Russian spacecraft.

Around five days before launch, the cosmonauts participate in a flag-raising ceremony outside the hotel and visit the cosmodrome's museum, where they sign books, postal covers and mementos.

Meanwhile, the Soyuz launcher is being assembled in a building near the pad, situated 35km (22 miles) north of the town. As the rocket lies on its side on a special railway wagon, the Soyuz spacecraft inside its protective white-painted metal shroud, with the escape tower on its nose, also rests on its side in a nearby cradle. The spacecraft and rocket are slowly moved together and firmly connected.

A locomotive pushes the wagon bearing the completed space vehicle out of the assembly building 48 hours before launch, which is normally arranged for 07:00 local time. The journey of about 4km takes place at walking speed and is often accompanied by officials, engineers, site workers and journalists strolling

TOP Soyuz TMA-15 astronauts Frank De Winne of Belgium (left) and Robert Thirsk of Canada (right) help their commander Roman Romanenko plant his commemorative tree at Baikonur. *(ESA)*

ABOVE Soyuz rocket stages in the assembly shed, with a Soyuz TMA spacecraft inside its white launch shroud, having the launch escape tower fitted.

RIGHT The Soyuz TMA-13 spacecraft is towed to the launch pad in October 2008. *(NASA-Bill Ingalls)*

for its space launches. Prior to the demise of the USSR in 1991, Russia and Kazakhstan were both part of the same country, the Soviet Union, so the distinction did not matter. Now, however, Russia wants its launch and recovery facilities to be on its own territory, which is why Vostochny is being constructed.

Baikonur's public spaces have a giant stone

alongside in an air of excited anticipation. Base personnel and guests lay coins on the rails to get flattened by the weight of the locomotive and the rocket wagon as a good luck charm.

The train arrives at *Gagarinskiy Start*, Gagarin Pad (or Gagarin Start), the historic launch pad from which the Sputnik and Luna satellites, almost all Soyuz vehicles and of course Yuri Gagarin himself, departed. Almost every Russian launch to the space stations Salyut, Mir and ISS, plus most of the Progress cargo freighters, have gone from this facility.

The rail line and services run on to the pad complex from the west side, because to achieve orbit the Soyuz must depart in a broadly north-easterly direction across vacant land. There is a giant flame pit on the east side, its steep western end straddled by the launch pad on two gigantic legs. Half a dozen tall towers for lightning conductors and floodlights surround the structure.

This horizontal assembly of the launch vehicle, and its erection upon arrival at the pad, is quite different from the approach adopted by the Americans for their Apollo lunar program. They assembled the Saturn V stage-by-stage in a vertical configuration inside the giant Vehicle Assembly Building (VAB), then it was driven to the launch pad fully upright, together with its gantry, on a crawler transporter. Some modern rockets, such as the SpaceX vehicles, are adopting the horizontal approach.

The cosmonauts awaiting launch do not observe the transport and erection of the Soyuz rocket on the launch pad, since that is considered to bring bad luck.

The crew also imitate some actions performed by Gagarin as he approached his pioneering mission. Two days before launch the crew get haircuts, or visit the hairdresser for a trim. On one of their last evenings before leaving Earth, they view the popular Russian action movie *White Sun of the Desert*. Although this was not released until 1970, it nevertheless has now become part of the ritual.

Launch day

Awakened very early on the day of launch, many cosmonauts have a very light breakfast. Canadian astronaut Chris Hadfield restricted himself to clear fluids and gruel, with

ABOVE The Soyuz FG rocket is supported by four gantry arms that secure it just above its four boosters, seen here before the launch of Soyuz TMA-15M in 2014. *(NASA)*

"sardonic awareness that we might see it all again in a few hours – post launch nausea is common". The crew, their spouses and officials sip a glass of champagne, or perhaps something lighter, and those who are bound for space sign the doors of their rooms at the Cosmonaut Hotel.

BELOW Canadian astronaut Chris Hadfield gets his pre-launch haircut at Baikonur. *(NASA)*

RIGHT In a launch day ritual, ESA astronaut Paolo Nespoli signs the door of his room at the Cosmonaut Hotel, Baikonur. (ESA)

BELOW British astronaut Tim Peake says goodbye to his young sons as he heads for the launch of Soyuz TMA-19M in December 2015.

RIGHT In another curious Russian launch tradition, modern male cosmonauts repeat Gagarin's toilet stop by the wheel of the bus as they travel to the launch pad. (Roskosmos)

Before leaving the hotel, the cosmonaut group sits down in silence for a few moments of reflection, to calm the mind. This is an old Russian tradition at the start of any long journey, and is said to help ensure that the trip will be completed safely. It is also a final opportunity to make sure nothing has been forgotten.

Nowadays, a Russian Orthodox priest blesses the crew in the hotel lobby, but during Gagarin's time, when the USSR was officially an atheist state, this was not practised. This tradition started in 1994 when commander Aleksandr Viktorenko sought a blessing for his Soyuz TM-20 crew, before setting off to the Mir space station. He flew with the third Russian female cosmonaut, Yelena Kondakova and German Ulf Merbold, who was the first ESA astronaut.

Next, two buses, the Star City bus with blue stripes and the Baikonur bus with yellow stripes, carry the cosmonauts and support personnel to the pad. Both buses are adorned with lucky horseshoes. The journey to the building where the crews don their spacesuits takes about 40 minutes.

Another curious ritual derives directly from the historic launch day of the world's first cosmonaut. As Gagarin's bus trundled towards the gantry embracing Vostok, he may have been a little nervous. He asked the bus driver to make an unscheduled stop, climbed down and urinated against the rear right wheel of the bus. One of his last acts on Earth became a tradition repeated by almost all subsequent cosmonauts.

Female cosmonauts are excused this tradition, but they have been known to bring along a container of urine to splash on the tyre. There has been only one known case of a male cosmonaut not participating in the ritual. That was Norman Thagard, the first American to fly on a Russian rocket, when heading for launch on board Soyuz TM-21 to the Mir space station in 1995.

OPPOSITE British astronaut Tim Peake gives a thumbs-up on 15 December 2015 at the launch pad for his Principia mission to the ISS aboard TMA-19M, with American crewmate Tim Kopra (middle) and Russian commander Yuri Malenchenko. (NASA/Joel Kowsky)

SUPERSTITION IN SPACE

Despite the rational and logical basis of science and engineering which underlies spaceflight, some human superstitions do appear to creep in from time to time.

Number 13 is not a traditional unlucky number in Russia, as it is in the West, where buildings often do not have a 13th floor, and passenger ships may not have a Deck 13.

Of the early American space programs, Mercury had only six flights and Gemini conveniently ended in 1966 with Gemini 12. The early Soviet Vostok and Voskhod programs never got beyond number six.

The first spacecraft to fly with the fateful number was Apollo 13, which launched to the Moon from the Kennedy Space Center on 11 April 1970 at 13:13 Houston time. On 13 April, as it approached the end of its outward flight, an explosion in the Service Module severely damaged the craft, and the crew of Jim Lovell, Fred Haise and Jack Swigert were fortunate to make it back to Earth alive.

Two years later, Soyuz 13 completed an uneventful mission as Valentin Lebedev and Pyotr Klimuk collected astronomical data using their on-board Orion 2 Space Observatory. The Soyuz program continued, with later variants of the craft carrying the same number without any apparent harm!

Soyuz T-13 was launched in 1985 and Soyuz TM-13 in 1991, with the latter proving to be the final launch of a Soviet spacecraft. The newly reconstituted state of Russia took over the former Soviet space program, starting with the landing of the (now-Russian) TM-13. Soyuz TMA-13 in 2008 was the 100th launch of a manned Soyuz spacecraft. Yet another variant ran through Soyuz TMA-13M without incident. The latest Soyuz model can be expected to reach number MS-13 in about 2019 or 2020.

Perhaps with Apollo 13 in mind, although some deny it, the US Space Shuttle flew with numbered missions from STS-1 to STS-9, then the scheme was changed to one in which the tenth flight of the program in 1984 was STS-41B, and the 13th was STS-41G. In 1988 NASA reverted to the original sequence, and the 26th flight was numbered STS-26R.

Bob Crippen, co-pilot on the first Shuttle mission, later said that James Beggs, who was the NASA Administrator at the time, was afflicted by *triskaidekaphobia*, the term for fear and avoidance of the number 13. Beggs stated privately there would not be a "Shuttle 13", and instructed that a new numbering system be devised. The unwieldy approach which emerged takes the last digit of the fiscal year in question (e.g. 4 for 1984), then 1 or 2 to indicate whether the launch pad was at Kennedy Space Center in Florida or Vandenberg Air Force Base in California (which was in the end never used), and a letter specifying the position of the flight in that

ABOVE NASA Administrator James M. Beggs decreed that there would be no Space Shuttle number 13. *(NASA)*

year's sequence – A, B, C and so on. This cryptic scheme attracted much criticism, so in 1988 James Fletcher, having taken over from Beggs, reverted to numbers and the 26th flight was numbered STS-26R (where the letter 'R' indicated that the mission plan had been revised).

Despite their healthy disdain for concern over the number 13, the Russians do have a few space superstitions. For instance, no space launch ever occurs on 24 October. On that day in 1960 a military rocket exploded on the pad at Baikonur and the fireball consumed over 70 people working around the vehicle, including Chief Marshal Mitrofan Nedelin, sitting on a chair exhorting his team to make urgent repairs.

Then, on the third anniversary of the Nedelin catastrophe, seven military personnel were also killed at Baikonur when a fire broke out in a missile silo. Rocket designer Boris Chertok later explained that after that incident, 24 October was considered bad luck at the firing range. It became a day off from work, and military testers even avoid undertaking substantial chores at home.

In another tradition, cosmonauts do not watch the roll-out and erection on the launch pad of their Soyuz space vehicle, as this is thought to bring bad luck.

As cosmonaut Aleksandr Lazutkin, who spent 185 days in space and had a narrow escape when a Progress freighter punctured the hull of the Mir space station, once said: "We may be materialists, but we are anxious to ensure that we get the support of forces unknown to us."

About three hours before launch, the three cosmonauts line up in front of the steaming rocket, officially report their readiness to fly the mission to their superiors, and salute. They then ascend a cramped elevator up the side of the gantry to a small metal cabin, from where they enter the spacecraft.

Because the Soyuz spacecraft was designed only to fly in space, where there is no air resistance, its shape is not streamlined – it is bulbous, and covered with thermal blankets, antennae, masts, solar panels and other variously shaped protrusions. During launch, the entire craft is covered by a white cylindrical protective shroud with a conical nose that is topped by the launch escape tower used for emergency separation from a malfunctioning launch vehicle.

In order to board the spacecraft, the cosmonauts have to pass one at a time through a square hatch in this shroud, immediately beyond which is a circular hatch on the side of the Orbital Module. This hatch can be used for an emergency spacewalk in orbit, but this has never yet been needed. From the Orbital Module, they then descend through a third hatch into the main compartment from which the craft is controlled, and where the cosmonauts sit for launch and landing. They enter about two-and-a-half hours before lift-off, with the commander taking the centre seat, and the hatches are closed.

Lift-off … or … Pusk!

As the countdown proceeds, the crew go through various checklists and wish each other good luck, or a happy landing. Finally the engines ignite gently for a few moments to build up to the required thrust, the Russian controller announces "*Pusk*!" (Launch!) and the rocket starts to rise.

"The Soyuz lift-off is quite soft," recalls American astronaut Leroy Chiao of his 2005 launch to the ISS on board Soyuz TMA-5. At first there is little sensation of movement, unlike on the US Space Shuttle, where the vehicle would flex dynamically upon ignition, rock back and rumble off the pad at a brisk pace. "On the Soyuz, the engines all start, then throttle very smoothly until the rocket gently rises off the pad. In fact you can't feel lift-off, and I only realised it

SOYUZ EMERGENCY ESCAPE

Like the US Mercury and Apollo launch vehicles, and newer capsules now being tested, Soyuz has a launch escape tower for emergency separation from a malfunctioning launch vehicle. This contains small, powerful rockets which can pull the Soyuz Descent and Orbital Modules clear of the rocket in the event of a serious mishap. This can occur either on the launch pad or during the early stages of flight. Square grid fins on the sides of the launch shroud are flipped down to provide aerodynamic control. Once a clean separation from the launch vehicle has occurred and the escape rocket burns out, the capsule is released to descend safely by parachute.

The Soyuz escape system has been used only once, in the so-called Soyuz T-10A launch anomaly of 1983. Cosmonauts Vladimir Titov and Gennadi Strekalov were hauled clear of their rocket several seconds before its intended launch, and the capsule landed by parachute 2.5 miles (4km) away. A fuel leak left the launch vehicle wreathed in flames. It exploded moments after the escape system operated, and the wreckage continued to burn for 20 hours.

BELOW Entering Soyuz – the rectangular hatch in the rocket shroud leads to a circular hatch in the Soyuz Orbital Module. From there, the cosmonauts climb down to their seats in the Descent Module.

ABOVE Pusk! The launch of Soyuz TMA-19M taking Yuri Malenchenko, Tim Peake and Tim Kopra on Expedition 46 to the ISS in December 2015. *(NASA)*

LEFT After a gentle lift-off, the Soyuz launch vehicle streaks into the sky. *(NASA/Joel Kowsky)*

when I heard the launch officer announce it, and I noticed the time on the clock."

After rising vertically, the launch vehicle soon performs a pitch-over manoeuvre to head in a north-easterly track, bound for orbit, and the acceleration forces on the crew increase to 1.5g. About 45 seconds after launch, at an altitude of about 7 miles (11km) and a velocity of 1,000mph (1,600kph), the rocket is subject to the maximum dynamic pressure from its passage through the dense lower atmosphere. The crew are now pressed into their seats with a force of 2g, but as the air thins out the launcher accelerates smoothly to a height of 25 miles (40km), where the four boosters and the escape tower are jettisoned. They fall back to Earth in an uninhabited area about 200 miles (350km) from the launch pad.

With the vehicle now so much lighter, and the core stage still firing at full thrust, the acceleration peaks and the cosmonauts experience 3.5g. British astronaut Helen Sharman recalls being intensely pressed back into her contoured seat.

Next, the shroud encapsulating the

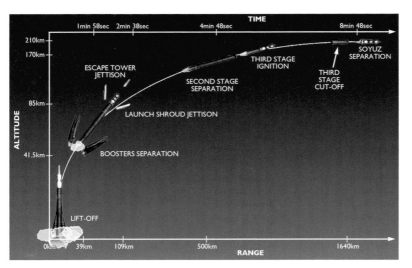

LEFT The Soyuz launch sequence from Baikonur to orbit. *(ESA/David Woods)*

BELOW Soyuz rocket boosters are jettisoned at over 25 miles (40km) altitude, photographed by an on-board camera. *(ESA/David Woods)*

spacecraft splits into two segments and falls away. This exposes the Soyuz spacecraft to the space environment, and if it is local day-time, the cosmonauts see light streaming into the cabin through the portholes.

About five minutes after launch, at an altitude of 105 miles (170km) and a velocity of about 8,200mph (13,200kph), the third stage ignites just as the thrust of the core stage begins to diminish. Then the cylindrical connection between the stages (which is clearly visible on the pad as a red band more than halfway up the vehicle) splits into three segments and tumbles away. The third stage burns for four minutes to more than double the speed and achieve orbit at an altitude of about 137 miles (220km).

Nine minutes after launch, the third stage is shut down and jettisoned. Like the spacecraft, the final rocket stage has also entered orbit. However, being larger and abandoned in an uncontrolled tumble, it will rapidly succumb to the drag from the tenuous atmosphere at this altitude and fall behind the Soyuz.

With the release of the acceleration pressure, Helen Sharman could finally turn and look out of the porthole. "We were still close to the Earth and I could see its curvature with black space above. It was so bright and vibrant I had to screw my eyes up to see. The sea was an amazing blue which I've never seen since."

The crew are now weightless and their mascot, which had hung like a pendulum from the control panel during the launch, now bobs around at random, while pens and notebooks float across the cabin. Various radio and navigation antennae swing out from their folded positions, the twin solar panels deploy from the sides of the Service Module, and Soyuz finally flies free into the black

ABOVE American astronaut Kate Rubins appears on a TV transmission from inside the cabin during the inaugural launch of the Soyuz MS variant in July 2016.

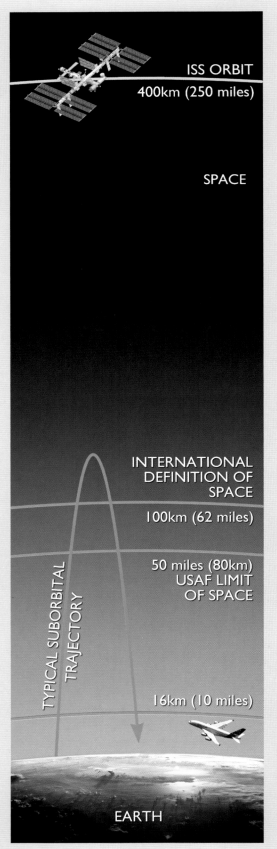

RIGHT The
atmosphere thins
out in layers to the
accepted boundary
of space at 62 miles
(100km) altitude.
(David Woods)

ISS ORBIT

400km (250 miles)

SPACE

INTERNATIONAL
DEFINITION OF
SPACE

100km (62 miles)

50 miles (80km)
USAF LIMIT
OF SPACE

TYPICAL SUBORBITAL TRAJECTORY

16km (10 miles)

EARTH

THE BOUNDARY OF SPACE

There is no absolute boundary to outer space above the Earth, as the atmosphere steadily thins out with altitude. The summit of Mount Everest, at 29,029ft (5.5 miles, 8,848m) above sea level, is the highest point on the planet, and human beings can survive there without breathing aids for only a few hours, due to cold and the lack of oxygen. The air pressure at the summit is around 4.9lb/sq in (0.34 bar) compared with 14.7lb/sq in (1 bar) at sea level.

Conventional jet airliners fly at up to 40,000ft (7.6 miles, 12km) and the now-retired supersonic airliner Concorde went to 60,000ft (11 miles, 18km). Just above that is the Armstrong Line, named after US Air Force physician Harry George Armstrong, at an altitude of around 62,000ft (19km) and a pressure of 0.9psi (0.06 bar). This is where all exposed human fluids such as saliva, tears, blood and the dampness within the lungs will boil off, hence survival outside of a pressurised cabin is totally impossible. No amount of oxygen supply can delay the rapid decline in human functions, loss of consciousness and death that rapidly follow.

The American U-2 and SR-71 Blackbird spy planes fly above this level. Their pilots wear pressure suits, as did the pilots of the X-15 research aircraft.

Helium-filled weather balloons and other unmanned high-altitude experimental balloons can readily attain 18 miles (29km), and one has gone up almost to 33 miles (53km). The record parachute jump from a balloon was done in this range in 2014, from nearly 26 miles (41km) altitude, by Alan Eustace wearing a pressure suit.

Above this altitude, meteors burn up due to atmospheric friction and leave their distinctive shooting star trail. They start to glow as soon as they touch the atmosphere, and are almost always burned to dust. Slow meteors come at speeds of 14 miles (23km) per second, and will make it down to 40 miles (70km) before vaporising. Much faster ones arrive at over 37 miles (60km) per second and are incinerated higher, at around 62 miles (100km) altitude.

The higher speed meteors travel at 133,000mph (214,000kph) or almost eight times the velocity of a satellite, hence, despite their usually tiny mass, they pose a very real threat of causing damage to a spacecraft or a spacewalking astronaut.

The Earth's polar aurorae, the Aurora Borealis and the Aurora Australis, also glow in the range of altitudes at which meteors burn up.

The generally accepted edge of space, 62 miles (100km) above sea level, is known as the Kármán Line. It was named after the Hungarian-American aerodynamic physicist Theodor von Kármán. He explained that in the atmosphere, a winged aircraft obtains its lift primarily as a result of travelling through the air, but in space, or the very tenuous outer layers of the atmosphere, an aircraft would have to travel so fast to generate lift that it would actually achieve escape velocity, making aerodynamic flight impossible above the line.

Above about 100km, aerodynamic controls are ineffective and a craft requires thrusters for attitude control, so the Kármán Line marks the approximate boundary between aerodynamics and astrodynamics.

BELOW Hungarian Theodor von Kármán defined the boundary of space.

void, the harsh yet startlingly beautiful environment for which it was designed.

Ahead lies a series of engine burns in a high-speed orbital chase to rendezvous with the International Space Station. A nominal rendezvous can now be achieved in about five hours, although previously it required 48 hours and in some cases a slow approach is still used.

After aligning with the selected docking port of the ISS, the Soyuz will automatically move slowly in to dock. As the probe on the nose of the spacecraft penetrates a conical receptor on the station, a mechanism engages to achieve a 'soft' docking. Then the two collars are drawn tightly together to form a rigid 'hard' docking. After a series of checks, the hatches are opened and the astronauts float into the station that will be their home for several days, weeks, months or, in a few exceptional cases, a year.

They are greeted warmly by the resident crew, who are pleased to see the colleagues they left on Earth several months earlier, and in some cases quietly grateful about the arrival of their taxi home.

ABOVE Soyuz TMA-17M just a few metres from its docking port at the International Space Station. (*Scott Kelly, NASA*)

BELOW ESA astronaut Tim Peake is greeted on arrival at the ISS by Mikhail Kornienko and Sergei Volkov.

If history taps the astronaut on the shoulder, he or she may get the opportunity to be the first to exit a spacecraft at a previously unvisited location. This could be on a physical body like an asteroid, a planetary moon or a new lunar landing site, or it could be making a spacewalk in Mars orbit or deep space beyond the Moon. The first words spoken will be recorded for posterity.

Yuri Gagarin, the world's first space traveller, did not have the opportunity to leave his craft, but as his Vostok rocket rose off the launch pad on a tower of flames, heading for space he calmly uttered the memorable phrase *"Poyekhali!"* meaning, "Let's go!"

This is what some famous astronauts said as they stepped from their craft:

BELOW Bruce McCandless makes the first untethered spacewalk using the MMU backpack from Shuttle STS-41B *Challenger* in 1984. *(Robert Gibson/NASA)*

Neil Armstrong, Apollo 11, upon first setting foot on the Moon's Sea of Tranquillity in July 1969.

"That's one small step for (a) man; one giant leap for mankind."

There has been some debate as to whether Armstrong actually uttered the 'a', but he later stated it was his intention to say it. If so, it is lost in communications static. NASA's *Apollo Lunar Surface Journal* records it as written above, with the 'a' in parentheses.

Charles 'Pete' Conrad, Apollo 12, the second Moon landing, Ocean of Storms in November 1969.

"Whoopie! Man, that may have been a small one for Neil, but that's a long one for me."

On the second lunar landing, Conrad had a bet with an Italian journalist who was convinced NASA would dictate his first words for him. He disproved this by uttering the quote he had previously agreed, which was a humorous reference to his own lack of stature. Conrad was one of the smallest astronauts, about 12cm shorter in height than Armstrong.

Bruce McCandless, STS-41B, the first untethered spacewalk with jet-powered back-pack, in Earth orbit, February 1984.

"Well, that may have been one small step for Neil, but it's a heck of a big leap for me."

Almost 15 years after Armstrong's historic saying, McCandless was not simply mimicking Conrad's humour. He had a special personal connection with the original Armstrong quote, having served as the CapCom in Mission Control Center in Houston who communicated with Apollo 11 for the first moonwalk. His was the first voice to acknowledge Armstrong's immortal words.

Recalling those events in 2014, McCandless said: "Part of my motivation had arisen from the fact that I had tried on several occasions prior to the Apollo 11 launch to get from Neil an inkling of what his 'first words' might be, so that I could be certain of getting them right. No luck. And, when he did utter them, the communications loop was just noisy enough that I couldn't tell whether or not he had uttered the phantom 'a'. Of course, all of the media representatives had to know *exactly* what Neil had said, and I was left 'holding the sack'."

THE ASTRONAUTS THAT WEREN'T...

Before Project Mercury put American astronauts into space in a small bell-shaped capsule on top of a Redstone rocket, other American pilots were already flying to the outer limits of the atmosphere.

The X-15 was a single-pilot rocket-powered aircraft which first flew in 1959, after being air-launched from beneath the wing of a B-52 carrier aircraft. It was the first true spaceplane, requiring small thrusters to control its orientation in the thin upper atmosphere where ailerons and rudders would not work properly.

During the X-15 program, which ran through most of the 1960s, 13 flights by eight different pilots exceeded an altitude of 50 miles (80km). This met the US Air Force's own criterion to qualify the USAF personnel among these pilots as being astronauts, but this assessment caused a number of discrepancies.

Firstly, the Air Force's definition of the limit of space does not correspond with the now generally accepted one, the Kármán Line, which lies at an altitude of 62 miles (100km).

Of the eight, five were USAF pilots who received the special *astronaut wings* badge. Among their number was Joe Engle, who later joined NASA and would fly two

Space Shuttle missions into orbit. But the other three pilots, not being USAF, were not issued astronaut wings.

Of the almost 200 X-15 flights, two, both flown by Joe Walker in 1963, exceeded 62 miles (100km) and therefore did qualify as true flights into space by FAI (Fédération Aéronautique Internationale) definition. But being one of the three non-USAF pilots, Walker's achievement was not immediately recognised with an award.

The three civilian X-15 pilots were eventually awarded astronaut wings in 2005, albeit according to the lower altitude definition initially used for their USAF counterparts. Walker, of course, qualified by the higher altitude FAI definition as well.

Incidentally, one of the civilian NASA X-15 pilots was Neil Armstrong, who would later become a legendary astronaut for greater deeds, but in his X-15 flights he reached neither the USAF nor FAI definitions of the boundary to space. He would not achieve this until flying Gemini VIII into orbit in 1966.

Nowadays, the US Federal Aviation Administration (FAA) also awards astronaut wings to non-NASA astronauts flying above the Kármán Line. The first two recipients were Mike Melvill and Brian Binnie for their sub-orbital flights in SpaceShipOne in 2004.

LEFT The X-15 rocket plane soared twice into sub-orbital space after release from its B-52 carrier aircraft. *(NASA)*

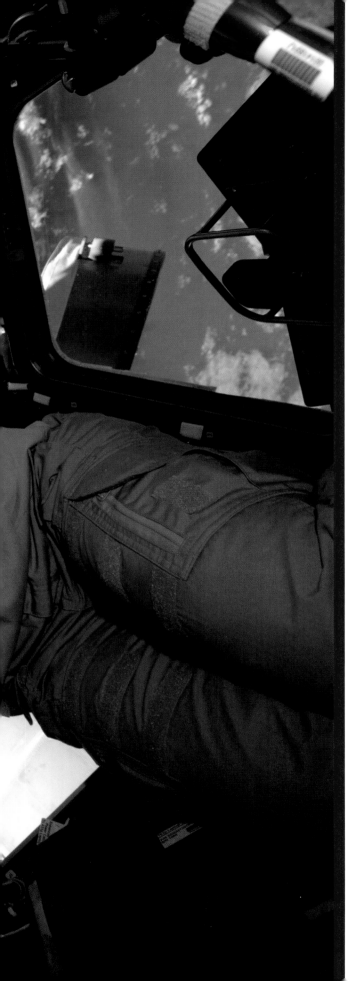

Chapter Seven

Life in orbit

A sojourn at the International Space Station is a high-intensity working trip crammed full of scientific experiments, maintenance tasks, and mundane house-keeping chores. There are 16 sunrises and sunsets each terrestrial day, punctuated by spectacular views of Earth and the cosmos. Occasional drama comes in the form of a spacewalk or the arrival of a cargo ship. All of this takes place in the extraordinary weightless environment of space.

OPPOSITE Samantha Cristoforetti photographs the Earth from the cupola aboard the International Space Station. This seven-window viewing unit is mounted on the Earth-facing side of the station and its optical-quality glass offers an all-round vista of the scene below. It is used for photography, science observation and, during time off, simply watching the planet roll by. *(ESA/NASA)*

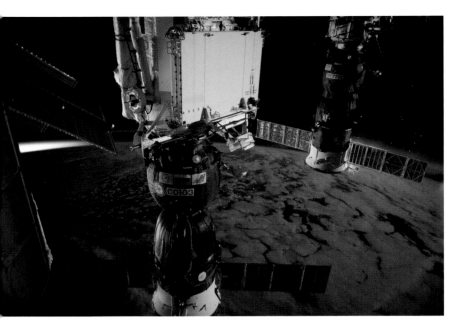

spacesuits since launch up until docking. Once safely docked, they can leave the cramped Descent Module with its seats and control panels, which they have occupied for the launch and subsequent manoeuvres.

They enter the much roomier Orbital Module, where they take off their constricting pressure suits. Although the OM when empty offers considerably more space to get changed, much of its volume at launch will usually be crammed with supplies and equipment destined for the ISS, so even here disrobing is something of a struggle. It also has the benefit of a small toilet.

However, the arriving astronauts can now enter the ISS in comfortable, loose-fitting flight overalls rather than clumsy pressure suits. Floating through the connecting hatch, they have a friendly reunion with the resident crew members, whom they know well from joint work before launch. They bring letters, spare parts for broken equipment and edible treats such as fresh fruit, bread and cakes.

They are now inside the biggest structure ever built in space, a permanently inhabited scientific base with a normal crew of six, which has been in orbit 250 miles (400km) above the Earth for about 20 years. The newcomers move their personal effects into their allocated accommodation berth and get settled in for their stay.

ABOVE A Soyuz (foreground) and a Progress supply craft are docked to the Russian segment of the ISS. Aurora flicker on the right below the stars, as the station races towards sunrise on the left.

After a journey of a few hours or two days, depending upon the type of rendezvous adopted by mission planners, the ferry craft bringing the crew to orbit docks to the International Space Station. China's Shenzhou ferry craft take the longer rendezvous option, arriving at the Tiangong space station a little under two days after launch.

If they have followed the rapid route to orbit in the Soyuz spacecraft, as is the norm these days, the crew will have been in their Sokol

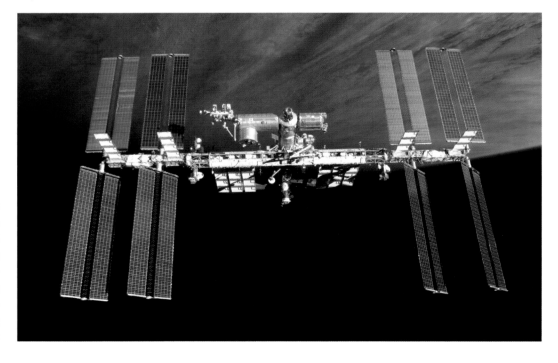

RIGHT The International Space Station in orbit 250 miles (400km) above the Earth, as seen from the departing Space Shuttle *Atlantis* on mission STS-129 in 2009. *(NASA)*

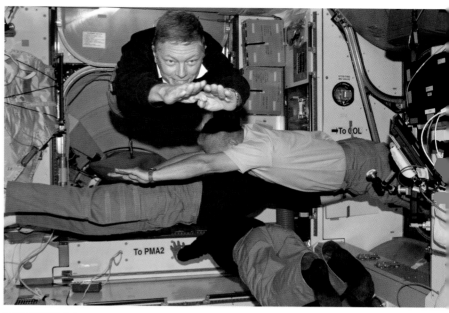

RIGHT Members of the joint STS-135/Expedition 28 crews performing floating exercises inside the ISS *Harmony* module. They are (from the top) Mike Fossum, Doug Hurley, Ron Garan and Satoshi Furukawa. *(NASA)*

Delights of weightlessness

For the first few days, they can finally enjoy the novel delight of unrestricted weightlessness, which was denied them in the congested space of the Soyuz ferry vehicle. They experience the extraordinary pleasure and freedom of floating free without touching the walls, doing slow somersaults and flying down tunnels with arms outstretched like Superman or Superwoman.

Canadian astronaut Chris Hadfield recalls: "It's like a gift, to be weightless all of a sudden. Take this wristwatch of mine – when I first got to orbit, I noticed it was floating and flying, almost like I had a living snake wrapped around my wrist. I'm going to keep my watch strap loose for the rest of my life, just because it reminds me of the magic of being weightless."

It is a sensory novelty too, for as they tumble, the Earthly sense of up and down is lost, and orientation has to be relearned. However, caution is initially advised in these entertaining acrobatics, as space sickness is a common consequence. The nausea usually wears off after the first few days so slow, deliberate motion is the best approach until astronauts get their 'space legs'. Vomiting in a confined cabin where the output disintegrates and floats off in all directions brings its obvious problems.

There is not much time for amusement, however, and a rigorous work regime is quickly established. Years of training and much money has been invested in getting the astronaut aboard the ISS, and time is precious. There are experiments to monitor, repairs to be made, and house-keeping chores to attend to – especially important in an unfamiliar environment where

RIGHT Gerald Carr, commander of the US Skylab 4 mission, jokingly demonstrates weight training in zero gravity as he balances astronaut William Pogue on his finger in 1974. *(NASA)*

unsecured objects, some almost irreplaceable, float away and may never be seen again.

With its extended solar panels, externally the ISS is about the size of a football field, while the interior is as large as a Boeing 747 aircraft. It has a habitable volume of 33,000ft^3 (934m^3), with several complex junctions and right-angle turns between modules, so losing things that refuse to stay where you left them is commonplace. The steady stream of circulating air from the environmental control system, and the weightlessness of objects, means things are continually on the move. The new astronaut soon learns to go to the air conditioning extractor vents in the hope of finding lost items. Velcro pads and elasticated cords adorn the walls to hold sundry items.

Other on-board activities include using one of the station's several robotic arms to perform work outside, with the astronaut controlling operations from the interior. There is the occasional spacewalk for repair work or scientific experiments, always undertaken in pairs, and non-spacewalkers have to assist with suit donning and doffing, communications, monitoring and support.

The arrival of an unmanned supply craft is a major event. Some like the Russian Progress craft, a modified non-recoverable Soyuz, can dock automatically, but the operation still needs to be carefully overseen, not least to avoid the sort of collision that once occurred (albeit

ABOVE Samantha
Cristoforetti working
on airway monitoring.
(ESA)

RIGHT Past and future
vehicles that visit the
International Space
Station, shown to
scale. *(Richard Kruse/
HistoricSpacecraft.com)*

RIGHT Scott Kelly harvested space-grown zinnias for Valentine's Day, 2016. The flowers were grown as part of the Veggie study that could lead to growing edible crops on long-duration missions.

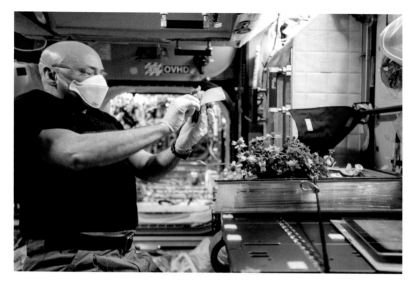

under remote manual control) at the Mir space station. That accident punctured the hull, and the cosmonauts hastily closed a hatch to the damaged module as their ears popped from the dropping pressure. The station was saved, but the depressurised *Spektr* module was a write-off.

The American commercially operated unmanned supply vessels (Dragon and Cygnus), and their Japanese counterpart (HTV Kounotori) come to a halt, station keeping some metres away from the ISS. They are grabbed by the robotic Canadarm2, under the control of the ISS crew, then manually guided to a berthing port on the *Harmony* module. The European ATV docks automatically like the Progress on the Russian segment of the station.

The vehicles are then unloaded and filled with discarded items for their return to Earth – destructively in the case of all except the Dragon, which can bring experiments and other equipment safely home.

In between stints of work, astronauts will often head for a window to take their fill of the unparalleled view endlessly unfolding below them.

ABOVE ESA astronaut Andre Kuipers, Expedition 30 flight engineer, in the Quest airlock during photo documentation of fluid and electrical interfaces.

LEFT ISS astronauts frequently head to the windows to take in the view. US astronaut Jeff Williams took this rare cloudless vista of Scotland, Northern Ireland, the north of England and the Isle of Man in 2016. *(NASA)*

RIGHT Crew-made meal of meat and cheese wrap, 2015.

Food and drink

Food is of course an essential part of everyday life on the ISS, but preparing and eating it brings its own peculiar set of challenges and hazards. Yuri Gagarin squeezed meat paste from a tube into his mouth. The Apollo astronauts ate carefully rationed bacon cubes, dehydrated potatoes, thermo-stabilised turkey and gravy, and shrink-wrapped peanut butter sandwiches. Fortunately, things have moved on since then.

Astronauts have long reported that food tastes different in microgravity and it is believed that this has something to do with fluids shifting to the upper body and the head. This can result in nasal congestion and a decrease in the perception of flavour, making many foodstuffs taste considerably more bland than usual. ISS crew members often crave spicy sauces and strong flavours to liven up their meals.

Food taken aboard the ISS aims to satisfy several key criteria, including nutritional appropriateness, ergonomics, weight at launch, shelf-life and practicality for eating in a weightless environment. How appetising food is perceived to be by the crew is also an important part of the food research activities of the various national space agencies. With nausea potentially only a rapid head-turn away, certain foods may be best avoided. One Space Shuttle commander banned any sort of fish from his flights as its overpowering smell could quickly swamp the entire craft – to the disappointment of a seafood-loving Japanese crew member.

'Cooking' is a somewhat euphemistic description of how the ISS crew prepare their meals. Much of the food can be eaten straight from the packets and drinks are dehydrated. Foods that can be rehydrated from a hot water spigot are efficient in several ways, because dehydrated food reduces the amount of weight to be carried into orbit, and it is instantly hot. Fresh food delivered by occasional arriving vehicles is a welcome addition to the pre-packaged diet.

Other desirable food attributes, along with quick and easy to serve, are being simple to clean up and leaving little waste. Something as innocuous as a stray breadcrumb can dry out into a combustible particle, float behind a panel, and short out sensitive electronic equipment.

Food is prepared on a tray or table top,

BELOW Expedition 5 crew members Peggy Whitson and Sergei Treschev make a barbeque cheeseburger and a grilled chicken sandwich aboard the ISS.

which has magnets, springs and Velcro to hold the cutlery and food packets in place. But chasing and gobbling up floating bite-sized food and globules of water is always popular entertainment.

Alcohol might generally be thought to be an unwise component of the space diet, but despite generally prohibitive regulations it has occasionally found its way aboard spacecraft.

Sherry, on account of its chemical stability, was considered and tested as part of a balanced diet for the nine American Skylab astronauts in 1973 and 1974, but the idea was abandoned due to adverse publicity. Earlier Buzz Aldrin, but not Neil Armstrong, had consumed a small quantity of communion wine on the Moon, but presumably not for its intoxicating effects.

The Russians seem to have shown a more liberated attitude to alcohol in space, at least in private. Cosmonaut Aleksandr Lazutkin, who flew to the Mir space station in Soyuz TM-25 in 1997, spending 185 days in orbit, has said that Russian cosmonauts were permitted cognac for extended missions. Photographs sometimes inadvertently show a glass brandy bottle floating in the Mir space station, or a can of Gösser Austrian lager beer strapped to the cabin wall. The first Bulgarian cosmonaut, Georgi Ivanov, reportedly resorted to brandy quite liberally when Soyuz 33 was forced to delay its return due to a malfunctioning engine.

Going to the toilet on the ISS, which has two facilities, Russian and American, is a relatively straightforward affair compared with the paraphernalia of sticky-rimmed fecal bags and overboard urine dumps that prevailed in an earlier era. Urine is collected in an airflow passing down a tube attached to a funnel. The end device is of different shapes for men and women, and each astronaut has their own attachment.

Gentle suction is also applied to a small toilet bowl where solid waste is expelled. On Earth, astronauts undergo 'positional training' to make sure solid waste goes directly into these space toilets, and that they do not miss the narrow opening. The training toilet has a camera at the bottom, but fortunately astronauts do not actually go to the bathroom during training. By watching a video screen in front of them, they can check that their alignment is accurate.

ABOVE ISS Expedition 34 crew member Chris Hadfield of Canada juggles some fresh tomatoes.

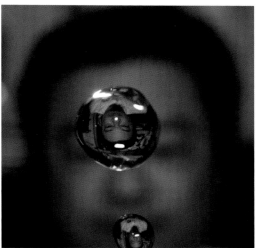

LEFT US astronaut Leroy Chiao's face is refracted in water globules.

LEFT British astronaut Tim Peake took Yorkshire tea with milk and sugar on his Principia mission to the ISS.

LEFT Astronaut Karen Nyberg performs an Ocular Health Fundoscope Examination on the ISS.

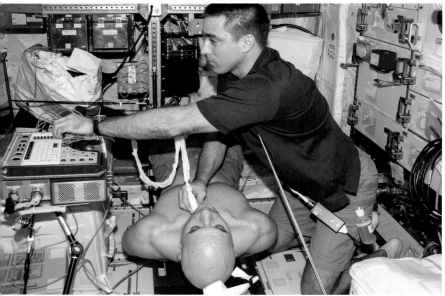

CENTRE NASA astronaut Chris Cassidy performs an ultrasound on ESA astronaut Luca Parmitano at the *Columbus* laboratory on the ISS. ISS scientists have carried out more than 1,000 studies so far.

Medical effects

With proper training and exercising techniques, learned from many early space missions at some cost to debilitated astronauts, the human body can adapt well to weightlessness. Some, but not all problems, can be avoided or delayed with the correct regime.

During the first few days of entering space, the human body has yet to adjust fully to weightlessness. The heart continues to pump blood to the head as if it were still countering the pull of gravity. Astronauts newly arrived in space have a puffy, bloated facial appearance for a few days, known as the 'moon face'. However, the cardiovascular system soon adapts to its new environment and provides a more regular distribution of blood throughout the body.

Longer-term weightlessness has more pronounced negative effects. These include muscles atrophying due to lack of use, resulting in lumbar stiffness and lower back pain, among other things. The skeleton deteriorates through calcium loss, in which astronauts shed on average more than 1 per cent of their bone mass for each month spent in space. Other medical effects include a slowing down of the cardiovascular system, reduced red blood cell production, loss of balance, deterioration of eyesight and weakened immune system, partly caused by radiation. To these systemic issues can be added some less serious irritations – nasal congestion, sleep disturbance and excessive flatulence.

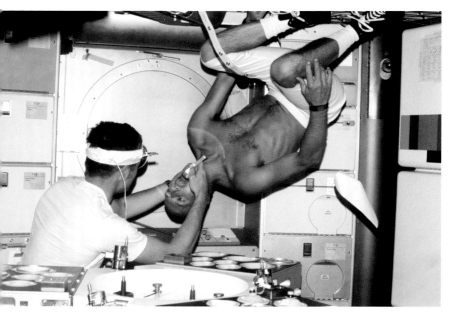

LEFT Dr Joseph Kerwin checks the teeth of his commander Pete Conrad on the Skylab 2 mission. In the absence of an examination chair, Conrad simply floats upside down. *(NASA)*

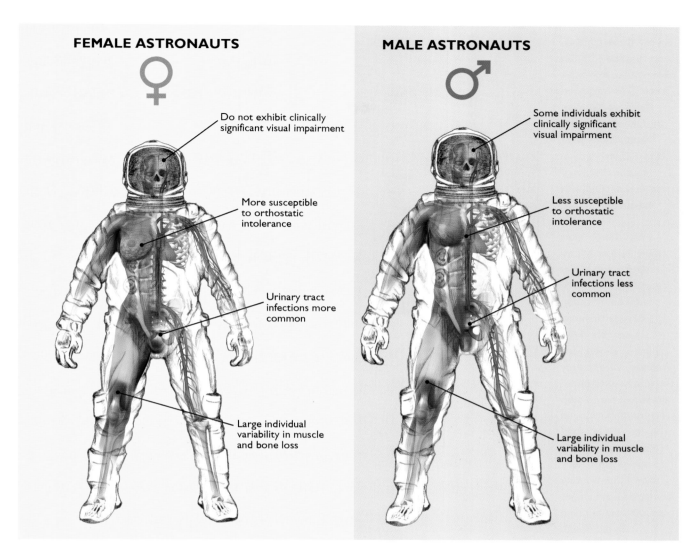

FEMALE ASTRONAUTS

Do not exhibit clinically significant visual impairment

More susceptible to orthostatic intolerance

Urinary tract infections more common

Large individual variability in muscle and bone loss

MALE ASTRONAUTS

Some individuals exhibit clinically significant visual impairment

Less susceptible to orthostatic intolerance

Urinary tract infections less common

Large individual variability in muscle and bone loss

Astronauts counter their slow deterioration with a strict daily exercise regime. They spend two hours a day on various exercise machines to stave off physical decline, and also to make it easier to re-adapt to normal gravity when they return to Earth.

There are several exercise machines on board the ISS for this purpose. They include two treadmills, a resistive exercise device, and a cycling machine, or ergometer. In each case, the astronauts have to be attached to the machine so that they will not float away.

The treadmills are used to simulate walking and running in normal gravity, and the astronaut is secured in place by elasticated cords connected to a harness which they wear. The cycle can be used to exercise either arms or legs. The resistive exercise device is somewhat like weightlifting on Earth, and it allows the

ABOVE Male and female bodies react differently to the space environment. NASA and the National Space Biomedical Research Institute published details of how spaceflight affects cardiovascular, immuno-logical, sensorimotor, musculoskeletal, reproductive and behavioural aspects of men and women.

LEFT US astronaut Sunita Williams exercises on the treadmill aboard the ISS.

RIGHT Steve Lindsey, Shuttle STS-133 commander, exercises using the advanced Resistive Exercise Device (aRED) in the *Tranquility* node of the ISS while space shuttle *Discovery* remains docked with the station. *(NASA)*

FAR RIGHT Scott Kelly (left) and Terry Virts (right) of ISS Expedition 43 are at work on a Carbon Dioxide Removal Assembly in the Japanese Experiment Module. Virts has his left foot hooked under a restraint bar to stop him drifting away, as he uses his notepad computer as a guide to his work. Kelly shows that wearing spectacles is no longer an obstacle to being an astronaut.

user to undertake a series of physical exercises while restrained by elastic bungee cords. A portable computer monitors heart rate and other vital signs while the astronauts are using the machines.

Cosmic radiation damages human cells and, though rare, large solar flares could inflict a fatal radiation dose in minutes. Crews on the Space Station are partially protected by the Earth's magnetic field, as the magnetosphere deflects the solar wind around the Earth and the ISS. Nevertheless, solar flares are powerful enough to warp and penetrate the magnetic defences,

and remain a hazard. In 2005, the crew of Expedition 10 took shelter as a precaution in a more heavily shielded part of the station designed for this purpose. Flying beyond the limited protection of Earth's magnetosphere, interplanetary crewed missions would be much more vulnerable, and radiation is a serious planning dilemma for more distant space flights.

These negative physiological effects are of varying durations, and start to reverse upon return to Earth. The body recovers fairly quickly, nevertheless it will take a few weeks to get back to feeling normal, a few months until you can run properly, and a matter of a few years to grow the skeleton completely back.

Time off in orbit

The Russians started the tradition of having a guitar aboard space vehicles, on the Mir space station, and Yuri Romanenko wrote 20 songs while living there in the late '80s. A guitar is now-permanent fixture on the ISS.

US astronaut Ron McNair brought his saxophone with him on Shuttle mission STS-41B in 1984, but the tape of that music was unfortunately recorded over.

Two US astronauts, Cady Coleman and Ellen Ochoa, brought flutes with them into orbit. In 2011, a recording of Coleman playing Bach's Bouree was blended with another from Ian Anderson of Jethro Tull, in the first Earth-space duet.

BELOW A SpaceX Dragon re-supply craft hovers off the ISS beyond the cupola windows, waiting to be grappled by the Canadarm2 on left.

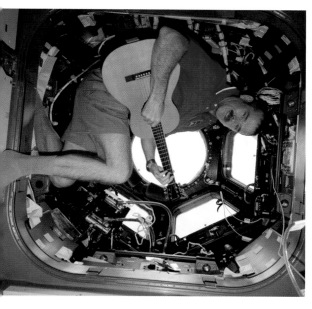

ABOVE Canadian astronaut Chris Hadfield with his guitar in the cupola. He made a music video floating inside the ISS of his cover version of David Bowie's *Space Oddity* in 2013.

ABOVE RIGHT Tim Peake reads Yuri Gagarin's *Road to the Stars* in the ISS cupola. This copy had previously been to the Russian Mir space station with the first British astronaut, Helen Sharman, in 1991.

RIGHT NASA astronaut Scott Kelly's personal living quarters on the International Space Station. Scott tweeted: "My bedroom aboard ISS. All the comforts of home. Well, most of them. #YearInSpace".

Coleman often played the flute in the Space Station's cupola, which she found to be the best venue because other modules have cloth bags and screens which absorb sound and offer poor acoustics. Being Irish-American, she also played a wooden flute, borrowed from Chieftains band member Matt Molloy, in the ISS on St Patrick's Day 2011.

Chris Hadfield famously videoed himself playing and singing David Bowie's *Space Oddity*

RIGHT The shape of the Moon as seen from the ISS is distorted by the Earth's atmosphere near to the horizon.

SPACE RECORDS

The Fédération Aéronautique Internationale (FAI), founded in 1905, is the world governing body for air sports. Although it is universally known by its French initials, its English name is The World Air Sports Federation. Based in Switzerland, its remit has nowadays been expanded to cover flights into space.

The FAI has the basic aim of furthering aeronautical and astronautical activities worldwide, and is recognised by the International Olympic Committee (IOC). It ratifies international air and space records, and stages competitions.

These include, for instance, human spaceflight duration records set on the ISS and various categories of spacewalking achievement.

The FAI Sporting Code of 2009 defines categories of achievement, and Section 8 covers astronautics. It recognises two types of what the FAI calls 'spaceships' – Class K, which covers manned *spacecraft*, and Class P for manned *aerospacecraft* that need to be capable of sustained and controlled flight in the atmosphere as well as in space, followed by a soft touchdown. There are three categories of space missions – sub-orbital and orbital for both Class K and P, and 'missions to celestial bodies' for Class K only.

The Sporting Code provides useful definitions relevant to space. For instance, a 'mission' comprises all the happenings and activities, scheduled or not, of a spaceship and its crew from the moment and place of take-off to the moment and place of termination of flight.

A reusable spaceship can be either Class K or Class P, but must be capable of making two manned consecutive flights such that a minimum of 90 per cent of the elements constituting the take-off empty mass of the first flight will be present in the take-off empty mass of the second flight.

Recognised spaceflight records include duration; altitude; distance from Earth; greatest mass lifted to altitude; extravehicular duration in space (that is, spacewalk); and accumulated spaceflight time (the total number of hours in space registered by one astronaut); and many more. Separate space records for women may be set in any of the categories.

There are several special conditions for spaceflights, including:

- All flights must exceed an altitude of 100km in order to qualify for records.
- A new world record must exceed the previous one by 5 per cent.
- The pilot and crew must be inside the spaceship component at take-off and all of them must reach the place of flight termination alive.
- The pilot and crew of an aerospacecraft shall remain inside the vehicle during descent and landing. For spacecraft, any method of descent and landing is acceptable, provided that it is described in detail in the pre-flight plan.
- For the purpose of all records (in particular for duration and distance records) the flight will start at the place and time of take-off and will finish in the manner defined above.
- A mission is sub-orbital if each of its arcs of trajectory above an altitude of 100km has a length of less than 40,000km (in the non-rotating geocentric set of axes).

The fourth condition above posed a problem for the very first human spaceflight because the descent method of the Vostok spacecraft required Yuri Gagarin to eject from his craft in the atmosphere as it descended, and to parachute to Earth separately. The Soviets did not reveal this at the time.

The rule had been designed with aircraft in mind – it could hardly be called a successful flight if the pilot bailed out over his destination and abandoned his aircraft to crash in an uncontrolled manner. When Gherman Titov flew on Vostok 2 several months after Gagarin, he revealed that he landed by parachute after leaving his capsule, which also descended smoothly under its own parachute. After analysis, the FAI wisely changed this rule for spaceflight to recognise the validity of Gagarin's flight. The fourth condition above for 'spacecraft' acknowledges this.

BELOW Logo of the Fédération Aéronautique Internationale which keeps spaceflight records. *(FAI)*

as he 'floated in a most peculiar way' through the modules of the Space Station. It has been viewed over 30 million times on YouTube.

Both Coleman and Hadfield played their music in the cupola, whose seven windows afford an unrivalled view of Earth. A relatively late addition to the ISS, for all the astronauts it has become one of the most popular spots aboard.

Cosmic vision

It is from here, and also to a lesser degree from the Space Station's other windows, that the most enduring sensory and even philosophical impact of space travel is experienced. Outside, the Earth rolls by continuously in a fascinating cavalcade of sights – clouds, oceans, islands, deserts, forests and cities. Because most of the viewing distance down to the surface of the Earth is through the vacuum of space, there is little to obscure the scene, which is crystal-clear and enables the astronaut to discern some extraordinary details.

Objects that are about one arcminute across are visible, which from the altitude of the ISS would be approximately 300 feet (100m), but with sufficient contrast smaller artefacts can be seen. Discernible individual structures include sports stadiums, oil tankers, highways and many large buildings. The often-quoted example of the Great Wall of China has a width that is below the theoretical resolution of the human eye, but that would change where there is differing land use on either side, or its height casts a long shadow at sunset or sunrise.

The early astronauts orbited at much lower altitude than the ISS. The early Gemini flights, for instance, had a perigee of around 100 miles (160km) and one astronaut saw the billowing trail of a truck being driven along a dusty desert road.

Borders are generally invisible, and this is one of the strongest impressions that astronauts of all nationalities get of the Earth – their own national patch of the planet is indistinguishable from neighbouring states. However, a few frontiers can be seen on account of contrasting land conditions on either side, including the Mexico-USA and Egypt-Israel borders.

Half of each orbit occurs in darkness, of course, because the Earth itself is blocking the light of the Sun and it is night-time below. Every 92 minutes a colourful sunrise is experienced, followed by an equally dazzling sunset 46 minutes later, as the station travels 25,000 miles (40,000km) around the Earth on every orbit, at 17,150mph (27,600kph). The atmosphere is seen for what it is, a thin and delicate sustaining layer overlying the planet, providing the most tenuous imaginable hold on life for its inhabitants.

The night-time view of Earth is as entrancing as the daytime one, displaying a network of city lights, roads, oil field flares and lightning storms. Air glow forms a luminous greenish band above the curved horizon, aurorae flicker at the poles, and the stars, seen without the normal atmospheric veil, are unimaginably numerous and bright.

Sights like these have fundamentally changed the perceptions of many space travellers, lasting long after they have returned to the frenetic hustle of Earthly life.

ABOVE From the ISS over northern Spain, looking across the Bay of Biscay at night to France, England and Ireland, as the Aurora Borealis play above.

BELOW Cloud-covered Italy and the Adriatic Sea roll past the Space Station window, photographed by Luca Parmitano. *(ESA)*

Chapter Eight

Walking in space

━━━●━━━━━━━━━━━━━━━━━━━

Today's astronaut must learn to walk in space, a complex skill which is essential to repair their spacecraft or to make an emergency escape. Spacesuits and handling techniques have evolved enormously since the alarming experiences of the early outings. Nowadays spacesuits are the pinnacle of protective technology, in effect being a miniature personal spacecraft, enabling the wearer to move and work comfortably through extremes of heat and cold, light and dark, in the vacuum of space.

OPPOSITE Robotic arms, as used on the Space Shuttle and the International Space Station, have become a useful means of anchoring and moving a spacewalking astronaut while they work. Here, Stephen Robinson, a mission specialist on Space Shuttle mission STS-114 *Discovery*, is perched at the end of Canadarm2. *(NASA)*

Only so many space activities can be undertaken from inside the spacecraft, be it flying, observing, manoeuvring, docking and the handling of exterior objects with the electromechanical manipulator arm. Eventually the astronaut encounters tasks which require him to venture outside, on what is popularly called a *spacewalk*.

Cosmonaut Alexei Leonov became the first person to undertake this dangerous activity, when he opened the outer hatch of the inflatable airlock on Voskhod 2 and pushed his weightless, spacesuited body out into the void.

The term spacewalk was not one used by Leonov. When his boss, the legendary Soviet spacecraft designer Sergei Korolyev, first explained the experiment, he used the somewhat better analogy of swimming rather than walking: "Now, in the same way that an Earthly sailor should be able to swim in the sea, you must learn how to swim in space."

The Americans repeated this feat two months later when Ed White exited the Gemini IV spacecraft. Unlike Voskhod, this had no airlock and White's commander, Jim McDivitt, remained inside the depressurised cabin to monitor White's spacewalk. In their techno-speak way, NASA engineers came up with a jargon term for the exploit: extra-vehicular activity. This was quickly contracted to one of the ungainly three-letter acronyms which permeate the entire US space program, an EVA.

These days, the technical term mostly used is EVA, referring both to a floating excursion outside the spacecraft in Earth orbit, and to a proper walk, or even a drive, on the surface of a body such as the Moon or Mars. The widely understood popular term remains spacewalk for the orbital version, and moonwalk and occasionally marswalk for the respective planetary surface excursions.

More than 4,000 hours of EVA have been conducted by individuals of 13 nationalities in Earth orbit. These have been undertaken from free-flying Voskhod, Gemini, Soyuz, Apollo, Shuttle and Shenzhou spacecraft, and from space stations Salyut, Skylab, Mir, and of course the present International Space Station (ISS). During their six lunar landings, the Apollo crews conducted 14 moonwalks and one observational EVA in which David Scott put his head out of the LM's upper hatch to obtain a better view of the surrounding area. Apollo crews also undertook three unusual spacewalks in deep space between Earth and the Moon in order to retrieve film from cameras installed outside the spacecraft.

The most extraordinary spacewalking records are, however, attributable to the ISS where this has become an almost routine occurrence. Overall, by early 2017 a total of almost 800 EVAs had been safely conducted by over 220 individuals of both sexes. During most of these spacewalks the tasks were to assemble, maintain, repair and operate the space station and its experiments.

Some highly specialised clothing and equipment is required to keep a human being safe from the lethal dangers that exist outside of the protective environment of a spacecraft, or inside the craft if pressure is lost.

BELOW Ed White made the first US spacewalk from Gemini IV in 1965. *(James McDivitt/NASA)*

Spacesuits

Outside the hull of a spacecraft is an extreme vacuum that is far more intense than any that can be created in a pressure chamber on Earth. The temperature ranges from minus 120°C (minus 184°F) in the shade to plus 120°C (248°F) in direct sunlight, and there are intense radiations of various types. If human skin were exposed to such a vacuum, all moisture, including water in the blood, would ultimately vaporise. Of course, this has never been demonstrated by a medical experiment, but it is believed that a person would deteriorate to a non-viable state in about a minute.

However, as pressure chamber accidents have shown, although a person rapidly loses consciousness, the body remains intact in a vacuum. It does not explode as some writers have dramatically suggested.

An additional danger in space is posed by micrometeorids, tiny grains of dust that travel at many tens of kilometres per second. If not stopped by the spacesuit's layer of Kevlar, the same material used in bulletproof vests, such particles would readily pass right through the human body causing serious damage.

To protect against these dangers, the early astronauts wore a single suit throughout shorter spaceflights which was never removed, and was based on the pressure garments worn by high-altitude jet pilots. Nowadays, national space programs tend to employ two types of spacesuits for different purposes at various stages of the typical space mission.

Launch Entry and Abort suits

First there is the Launch, Entry and Abort (LEA) suit, designed to be worn inside the spacecraft for critical phases of the flight, such as launch and re-entry, where there is a risk of loss of cabin pressurisation or a release into the cabin of noxious gases. This lightweight suit will afford full protection in a vacuum inside the spacecraft during an emergency. It incorporates communications systems, helps to counter the dynamic loads of launch and landing, and has flotation facilities for use in the event of rescue from water.

LEA suits have been standard equipment since some risky suit-less flights by Soviet cosmonauts in the 1960s culminated in the depressurisation disaster of Soyuz 11, which exposed the three crewmen, Georgi Dobrovolsky, Vladislav Volkov and Viktor Patsayev to the vacuum of space. When the

ABOVE Yuri Gagarin at the launch pad in his SK-1 spacesuit.

Vostok capsule rebranded as Voskhod, the cosmonauts wore ordinary clothes. But pressure suits were essential for the follow-on Voskhod mission on which Alexei Leonov pioneered spacewalking. Then it was back to ordinary clothes for Soyuz missions until the Soyuz 11 disaster led to the development of a lightweight suit.

All Mercury, Gemini and Apollo astronauts wore pressure suits for launch, although they did not always wear them for the return to Earth. After the early tests in which pressure suits were worn, Space Shuttle astronauts wore ordinary clothes and clamshell-style helmets. Only after the loss of Challenger in 1986 did NASA reintroduce pressure suits.

The first modern American LEA suit, called the Launch Entry Suit (LES), was worn by Shuttle crews from 1988 to 1994, when it was replaced by the similar-looking Advanced Crew Escape Suit (ACES). Both included a parachute and flotation device and were individually sized, although some of them could be worn by astronauts of different heights. The LES and the ACES were known informally by the astronauts as 'pumpkin suits', owing to the colourful similarity to the popular North American fruit.

Modern cosmonauts and astronauts flying on the Russian Soyuz to the ISS wear the white Sokol (Falcon) suit which serves the same purpose as ACES during launch and landing, and predates it, having been used since 1973. Their

capsule landed automatically and the recovery team opened the hatch they were shocked to discover the crew lifeless. Their lives would have been saved by LEA suits.

Suits of this type include the orange spacesuits worn by Yuri Gagarin and the other cosmonauts of the Soviet Vostok program, and the similar-looking orange suits worn by most US Shuttle astronauts as they walked out to board the bus on launch day.

The suit for male Vostok cosmonauts was the SK-1 (Skafandr Kosmicheskiy). Valentina Tereshkova, the first woman in space, wore the modified SK-2. When it was decided to squeeze three men into a modified

RIGHT The crew of Shuttle mission STS-126, led by commander Chris Ferguson, depart for the launch pad. The sleeves of their tube-lined thermal suits are visible beneath their orange launch and entry suits.
(NASA/Kim Shiflett)

Chinese counterparts wear a similar-looking suit of domestic manufacture, reverse-engineered from Sokol suits purchased from Russia.

The emerging space tourism industry is also seeing the development of a new range of spacesuits, in some cases with a stylish design, to protect the amateur space traveller in the unlikely event of cabin depressurisation on their brief flight.

EVA suits

The second type is the full EVA suit for use in the harsh external space environment. It provides full protection from vacuum, radiation, the extreme variation in temperature, micrometeoroid impacts and the risk of damage from sharp objects that the astronaut might encounter while outside of the spacecraft. This suit is almost always either white or silver in order to reflect the intense unfiltered sunlight of open space.

Full EVA suits include the current American Extravehicular Mobility Unit (EMU) made for use on the ISS by Hamilton Sustrand, the Apollo spacesuit used by moonwalkers in the 1960s and '70s, and its successor used on the American Skylab space station.

There were also earlier suits developed by the USSR and USA for use in the Voskhod, Gemini and early Soyuz flight programs. The Orlan (Sea Eagle) suit, incorporating a novel

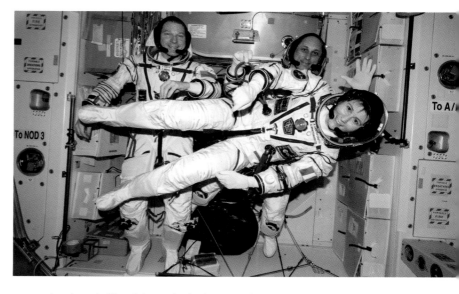

rear-entry door, is Russia's equivalent garment for EVAs on the ISS, and has also been used by American astronauts working outside with their Russian colleagues. The Chinese borrowed technology from the Orlan suit to produce its Feitian EVA suit, whose name means something like 'Sky Flying'.

Some suits have been capable of being converted from one type to the other, such as the American Gemini G4C spacesuit. The basic five-layer suit was designed to protect the astronaut from vacuum inside a depressurised vehicle. Another two layers were added to upgrade it to an EVA suit for nine Gemini spacewalks. For one spacewalk, the suit

ABOVE The crew of ISS Expedition 42/43 Samantha Cristoforetti (ESA), Terry Virts (NASA) (left) and Anton Shkaplerov (Roskosmos) show off their Russian Sokol spacesuits. *(NASA)*

LEFT Russian cosmonaut Fyodor Yurchikhin (left) enters his Orlan suit, assisted by Alexander Misurkin, who is already dressed for EVA.

received additional thermal protection against the efflux from a jet backpack that was intended to be tested, but in the event never was.

Astronauts flying the American Orion spacecraft use a modified ACES, called MACES, which can double as an EVA suit. This suit provides greater mobility than the original Space Shuttle LES and ACES, and employs a closed-loop system to preserve consumables. NASA has successfully evaluated its use on EVAs, such as during the Asteroid Redirect Mission (ARM) proposed for the mid-2020s.

Early EVA experiences

The first EVA by Leonov was completed successfully despite the primitive nature of his equipment and the great danger to which he was exposed by badly thought-out procedures. It proved such a salutary experience that the Russians did not attempt EVA again until they had developed a new suit and spacecraft more than four years later when two cosmonauts, Alexei Eliseev and Yevgeny Khrunov, spacewalked between two Soyuz craft.

Leonov had great difficulty re-entering the

airlock of Voskhod, because his poorly designed Berkut (Golden Eagle) spacesuit ballooned more than expected in the vacuum of space, causing his hands to retract from his gloves so he could not grip properly. Worse, the suit itself became rigid and he had to bleed off oxygen to reduce the pressure, risking the 'bends', a painful ache which afflicts deep sea divers if they ascend too rapidly. With extreme exertion, Leonov barely managed to haul himself back into the airlock, ending up exhausted and drenched in sweat. He had closely escaped being stranded outside, and inevitable death.

The USA resolved to build on their own first spacewalk, but it was a long and experimental process lasting the remainder of the Gemini program. Due to a lack of appreciation of space dynamics and inadequate training, many early EVAs had to be called off. The Americans were attempting to learn how to work in the weightless environment, performing various tasks such as using manual tools, cables and powered devices, and trying out manoeuvrable backpacks.

One of the main problems encountered was the inability to gain leverage in order to move or turn handles and screws. Unless a spacewalker is securely braced, when he tries to rotate a handle his entire body will turn in the opposite direction. Lacking an anchor, any attempt to recover his original posture inevitably makes things worse. Anything set in motion in the weightless environment, including an astronaut, will continue moving until stopped. It is all too easy to drift out of position, while tumbling in several axes. To restore stability, each of these distinct components of motion must be cancelled out.

However, this task is verging on the impossible when sensory overload is considered. The astronaut's work station, the spacecraft, the Earth, and the glaring Sun are all now crossing the faceplate of their helmet in a bewildering spectacle of shining metal, the blackness of space and the patterned Earth below. Furthermore, every 45 minutes the entire scene is plunged into darkness or dazzling daylight as the spacewalker and his craft enter or exit the Earth's shadow as a result of their incessant orbital motion. Lights need to be switched on or off, and even inside the insulated spacesuit temperatures rise and fall abruptly.

BELOW Chinese astronaut Zhai Zhigang exits Shenzhou 7 in a Feitian suit to make China's first spacewalk in 2008. His colleague Liu Boming is seen in the hatch entrance wearing a Russian Orlan suit. *(CCTV)*

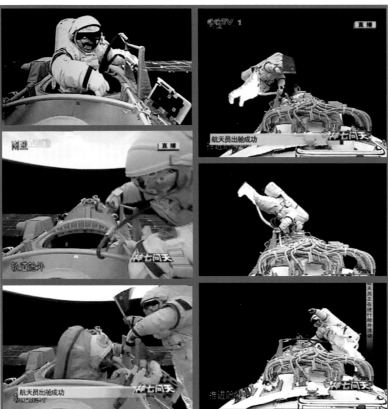

These factors, plus poor training, plagued the early Gemini spacewalks. The astronauts found they could not perform even simple tasks without tumbling out of control, as their exertions accelerated their metabolic rate and fogged up the insides of the helmet visors, seriously obscuring their vision. The reluctance of the highly competitive pioneer spacewalkers to give up resulted in overheating, extreme perspiration, and even weight loss.

Gene Cernan, who would later become the last man on the Moon, attempted a very complex EVA on Gemini IX-A in 1966, where he crawled to the rear of the craft to strap on an innovative jet-powered backpack. It all went seriously awry and, completely exhausted with a dangerously racing heartbeat, he was ordered back inside by his commander Tom Stafford. With the cabin re-pressurised, Stafford pulled open Cernan's visor to reveal his bright pink, overheated face, and was so concerned that he broke the rule about handling liquids in weightlessness and sprayed him with water. During his brief excursion Cernan had sweated out about 2 pints (1 litre) of water. This was found sloshing around in his boots after they landed back on Earth.

Towards the end of the Gemini program, the Americans diagnosed the difficulties of spacewalking and worked out procedures to deal with them. Unlike his predecessors, Buzz Aldrin rehearsed his spacewalks underwater, devising solutions to previous EVA problems, and when he flew the final Gemini mission in 1966 he was able to work effectively. Across three EVAs he set a record total of 5hr 30min of EVA time.

Overall, five Gemini space missions conducted nine EVAs, for a total of 12hr 22min outside. Their exploits laid a solid foundation of experience for Apollo. In particular, to preclude overheating, it was decided that moonwalkers would wear an undergarment with a network of thin tubes which would circulate water near the skin. There were appropriate heat exchangers in the life-support backpack.

The USSR resumed spacewalking in late 1969, when two manned Soyuz spacecraft docked in orbit and cosmonauts Alexei Yeliseyev and Yevgeny Khrunov, wearing the new Yastreb (Hawk) spacesuit, transferred from

Soyuz 5 to Soyuz 4, and returned to Earth in a different craft from the one in which they launched. Both spacewalkers experienced problems with the Yastreb, and like Leonov's Berkut, the suit was never used again.

Apollo EVAs

With the Apollo program came not just the requirement to conduct weightless floating EVAs, but also the need to walk on the Moon. Apollo 9 tried out the new Apollo spacesuit for the first time, when David Scott and Rusty Schweickart opened the hatches of the docked Command and Lunar Modules respectively in Earth orbit, and pronounced them highly functional. Unlike previous American spacewalkers, Schweickart wore a portable life support system (or PLSS, pronounced 'Pliss') in the style of a backpack that supplied consumables of oxygen and water. This

ABOVE Gene Cernan closes his eyes in exhaustion as he wrestles with the difficulty of working in weightlessness at the rear of his Gemini IX spacecraft in 1966. *(Douglas Cooper)*

RIGHT The Apollo A7L
EVA suit was used by
12 individuals to walk
on the Moon.
(Paul Calle)

was intended for later use by moonwalkers. Scott's suit, in contrast, was plugged into the spacecraft to receive consumables via an umbilical line.

By now EVA training was being conducted in a water tank in Houston where neutral buoyancy mimicked the sensation of weightlessness. This greatly improved skills and reduced the risks of spacewalking by familiarising trainee astronauts with the floating sensation and the action-and-reaction effect which had confounded the efforts of the first spacewalkers.

The Apollo program conducted many EVAs under the lunar surface conditions of one-sixth Earth's gravity and also in weightlessness. Although it caused some issues with mobility and dexterity, and its zippers became clogged with lunar dust, the Apollo spacesuit was a great success and many of its design features were retained by later suits.

The Skylab program, which came immediately after the Apollo lunar missions, used an Apollo-style suit which drew breathing oxygen and coolant from an umbilical that also carried communications and served as a safety tether. Typically two astronauts would go outside at any one time, while the third monitored activities from inside. A total of ten EVAs, amounting to over 82 hours, were performed during three missions to the space station.

The Russians established a steady lead in orbital spacewalking during the late 1970s, '80s

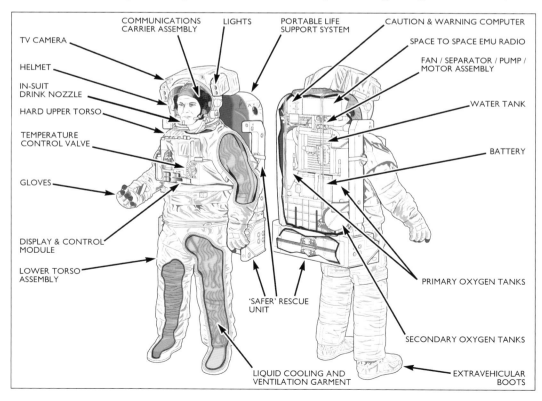

RIGHT The American
EMU represents the
pinnacle of spacesuit
technology.
(Ian Moores/
David Woods)

and '90s from the Salyut 6, Salyut 7 and Mir space stations. Their Orlan suit was first tested in 1977, and with various upgrades is still used on the ISS today.

The Space Shuttle program of the 1980s required an entirely new spacesuit, designed for frequent use by variously sized individuals, which would afford greater mobility and use a backpack life support system rather than an umbilical.

The Shuttle program pioneered a wide range of innovative EVAs, including the first untethered jet-pack-propelled spacewalks, capturing satellites manually, working in the Shuttle's large payload bay to repair satellites and the Hubble Space Telescope, repairing the Shuttle's exterior, and assembling the various elements of the International Space Station. More than 200 separate EVAs contributed thousands of hours of spacewalking experience.

American EMU

The Extravehicular Mobility Unit (EMU) represents the current pinnacle of spacesuit technology. It provides the astronaut with a completely independent system to survive, move around, work and communicate outside of a spacecraft.

It contains enough consumables to support an EVA lasting up to seven hours. This standard total is based on the following assumed time breakdown:

■ 15 minutes for egress of the main spacecraft (usually the ISS)
■ 6 hours of useful outside work
■ 15 minutes for ingress
■ 30 minutes in reserve.

The EMU is made up of two separate component assemblies designed to operate together as an integrated unit – the Space Suit Assembly (SSA) which is the garment itself, and the Primary Life Support Subsystem (PLSS), the backpack which provides the means to sustain the life of the suit occupant.

Space Suit Assembly – inner
The inner layers of the SSA which are closest to the astronaut's skin include the following elements:

■ Thermal Comfort Undergarment (TCU)
■ Liquid Cooling and Ventilation Garment (LCVG)
■ Operational Bio-instrumentation System (electrocardiogram)
■ Communications Carrier Assembly (CCA)
■ Disposable In-suit Drink Bag (DIDB)
■ Maximum Absorption Garment (MAG).

Overheating has always been a problem during spacewalks, on account of the exertion involved, the heat of the unfiltered sunlight falling on the outside of the suit, and the undesirability (and indeed danger) of profuse perspiration. Hence the liquid cooling garments are a vital component first introduced in the Apollo era in response to the overheating experienced with the Gemini G4C suit.

Immediately next to the skin, the astronaut wears a Thermal Comfort Undergarment. On top of this is worn the Liquid Cooling and Ventilation Garment (LCVG). This all-over outfit incorporates about 330ft (100m) of flexible, narrow tubing through which chilled water flows to carry away the heat generated by the body, which could not otherwise escape from the insulated outer suit, and helps maintain a

BELOW Astronaut Peter Wisoff wears the EMU while standing on a Portable Foot Restraint attached to the Shuttle's robot arm on STS-57 in 1993. *(NASA)*

constant internal temperature. Astronauts report it is a very odd sensation to feel one's body at a comfortable temperature despite considerable exertion, yet the head remains hot.

The Operational Bio-instrumentation System consists of three chest electrodes, a signal conditioner, and connecting cables. It provides an electrocardiogram (EKG or ECG) signal during the EVA, to check for problems with the electrical activity of the astronaut's heart.

The Communications Carrier Assembly (CCA) is basically a soft hat or helmet containing earphones and microphones so that the suited astronaut can communicate with spacewalking colleagues and those aboard the host spacecraft, as well as with Mission Control on Earth. It also presents caution and warning tones to the astronaut. The CCA is coloured white and black or brown, and has long been called the *Snoopy Cap* on account of its similarity in appearance to the white head and black ears of the dog in Charles M. Schulz's *Peanuts* comic strip.

The Disposable In-suit Drink Bag (DIDB) is a transparent bladder fixed inside the spacesuit with a protruding nozzle to enable an astronaut to drink water while spending up to seven hours in the suit without nourishment. NASA decided to switch to disposable bags starting with Shuttle STS-92.

The Maximum Absorption Garment is in fact a nappy or diaper used to absorb bodily waste in the event of the astronaut requiring to go to the toilet during a spacewalk, a function which they are unable to accomplish in the normal way.

Space Suit Assembly – Outer

The outer layers of the SSA comprise the following elements:

■ Hard Upper Torso (HUT) and arms
■ Lower Torso Assembly (LTA) and boots
■ Extravehicular (EV) gloves
■ Helmet and Extra-Vehicular Visor Assembly (EVVA).

Together, these provide the protective outer layer which shields the spacewalker from the harsh environment of space.

The Hard Upper Torso (HUT) is a rigid, protective fibreglass unit which fits over the upper body of the astronaut, and has attachment points for the separate arm units and the LTU. It houses the connection points for the life-support tubes that circulate water and provide oxygen. The Display and Control Module is also attached here. The HUT comes in three sizes, and the arm lengths can be adjusted by adding or subtracting rings. The DIDB is installed inside the HUT.

The Lower Torso Assembly (LTA) includes the lower half of the waist seal, suit trousers (pants) and treadless boots. A metal body-seal closure connects the LTA to the HUT. A moveable waist bearing assists the astronaut to turn left and right. The waist area has D-rings to attach tethers, short cords that clip astronauts to the spacecraft so they will not float away. Some spacesuits are plain white, while others have red rings or candy stripe dashed rings around the thigh area, to distinguish one spacewalker from another.

The EV gloves have to strike a balance between protecting the astronaut's hands from the space environment and any parts of the spacecraft that they touch, and providing sufficient flexibility to pick things up, turn handles and manipulate tools. Fingers are the parts of the body that become coldest in space, so the gloves have heaters inside the hardened tips. A rotating bearing connects the glove to the sleeve, allowing the wrist to turn.

The Helmet and Extra-Vehicular Visor Assembly (EVVA) is a transparent plastic bubble that locks on to the neck ring of the HUT, with its protective covering. The helmet has a vent pad that directs the flow of oxygen from the

BELOW The EMU spacesuit's EV glove has heaters in the fingertips.

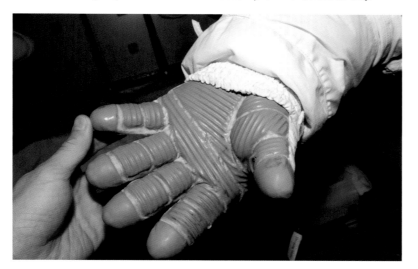

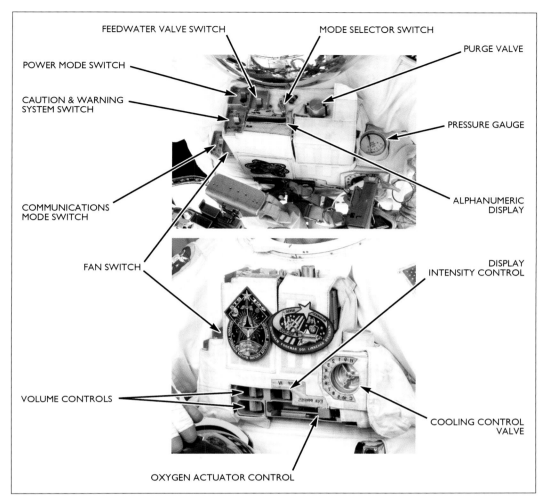

FEEDWATER VALVE SWITCH

MODE SELECTOR SWITCH

PURGE VALVE

POWER MODE SWITCH

CAUTION & WARNING
SYSTEM SWITCH

PRESSURE GAUGE

COMMUNICATIONS
MODE SWITCH

ALPHANUMERIC
DISPLAY

FAN SWITCH

DISPLAY
INTENSITY CONTROL

VOLUME CONTROLS

COOLING CONTROL
VALVE

OXYGEN ACTUATOR CONTROL

**LEFT The Display
and Control Module
is mounted on the
chest of the EMU.
Some labels are
written in mirror text
to be read in an arm-
mounted reflector.**
(NASA/David Woods)

Primary Life Support Subsystem into the front of the helmet, for breathing. The bubble is covered by the Extra-Vehicular Visor Assembly, which protects the spacewalker from extreme temperatures and the impact of small items. The reflective visor is coated with a thin layer of gold that filters out the bright, harmful rays of the Sun. A TV camera and lights are attached to the sides of the helmet, illuminating the scene in front of the astronaut for the half of each orbit which is in darkness, and conveying what he is seeing to Mission Control.

A number of other items are attached to the exterior of the suit. The main one is the chest-mounted Display and Control Module (DCM) which carries controls for the life support systems and various status displays, including messages from the Caution and Warning System (CWS). Because the suit occupant cannot see the front of the DCM, there is a mirror on one sleeve to enable the astronaut to read the settings, whose

labels are written in reverse. There is also a Cuff Checklist of tasks to be performed during the EVA. Introduced on Apollo 12, this was a small book with flip-pages and therefore the amount of information was limited, but nowadays it is an electronic device and spacewalkers can call up proper technical manuals.

Primary Life Support Subsystem

The Primary Life Support Subsystem (PLSS) is worn as a backpack, and provides the consumables to sustain the astronaut's life while wearing the spacesuit. It has tanks of breathable oxygen and a fan to circulate it, together with a device to remove exhaled carbon dioxide. The PLSS also holds water-cooling equipment to supply the LCVG under-garment. A battery provides electrical power for various items, including two-way radio. In the

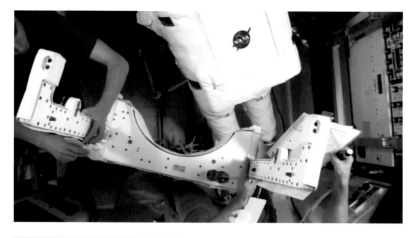

LEFT The SAFER unit ready to be fixed around the PLSS backpack on UK astronaut Tim Peake's Principia mission in January 2016. *(NASA TV)*

CENTRE German ESA astronaut Alexander Gerst is wearing his thermal cooling garment over his comfort undergarment, as he prepares to don the EMU in the ISS Quest airlock for a seven-hour EVA in 2014. His fellow spacewalker Reid Wiseman will wear the suit on the right. *(ESA)*

event of a problem, the caution and warning system in the backpack feeds data to the DCM to inform the user.

Fitted around the base and sides of the PLSS is the Simplified Aid For EVA Rescue (SAFER). This is a slimmer version of jet-packs used in the Space Shuttle era to enable astronauts to move themselves around in space without a tether. It is for use only in an emergency, when the tether being used to connect the astronaut to the space station comes loose and he floats away. Once its controls have been deployed, a small joystick will enable the SAFER's nitrogen-jet thrusters to return the astronaut to the station, in order to reconnect the tether.

With a few modifications to extend its operational life, the EMU is now the basic spacesuit of the American and European segment of the ISS, just as the latest version of the Orlan is on the Russian segment.

Conducting an EVA

EVAs are undertaken by pairs of astronauts for maximum safety and work efficiency. They help one another into their suits in the ISS Joint Airlock, named Quest, which can support both the American EMU and the modern version of the Russian EVA suit, the Orlan-MK. They first put on the MAG, or adult nappy, then cotton long johns and vest (TCU) to absorb any

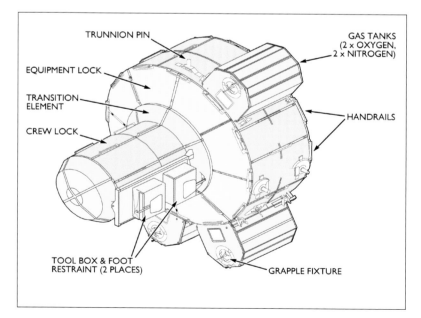

TRUNNION PIN

EQUIPMENT LOCK

TRANSITION ELEMENT

CREW LOCK

GAS TANKS (2 x OXYGEN, 2 x NITROGEN)

HANDRAILS

TOOL BOX & FOOT RESTRAINT (2 PLACES)

GRAPPLE FIXTURE

LEFT The Quest airlock on the ISS, from which all EVAs wearing the American EMU are conducted. The exit hatch is on the underside of the Crew Lock as viewed from this angle. *(NASA/ David Woods)*

sweat, which is overlain by the liquid cooling garment (LCVG). Next comes the Snoopy Cap with its communications set and the LTU. Finally the HUT, helmet and gloves are added. This takes about an hour.

Here they also don their backpacks and attach their tool selection to a mini workstation on the front of the suit. To adjust their bodies from the normal Earth atmosphere within the station to the spacesuit ventilation regime, they must pre-breathe with 100 per cent oxygen.

They enter the crew airlock unit, floating side-by-side and head-to-toe, then close the inner door and depressurise the chamber to a vacuum over a period of about half an hour. Then the outer hatch is opened inwards to reveal the flimsy outer thermal cover, which is opened outwards, and the way to open space is clear. With all their training behind them, they can finally float outside and see the stunning spectacle of the multi-coloured Earth and the blackness of space in all its magnificence.

Canadian astronaut Chris Hadfield explains his reactions to his first spacewalk in 2001: "I'd known I was venturing out into space, yet still the sight of it shocked me profoundly… Of course, I'd peered out the Shuttle windows at the world, but I understood now that I hadn't seen it, not really…[now] I could truly see the astonishing beauty of our planet, the infinite textures and colours. On the other side of me [was] the black velvet bucket of

ABOVE US astronaut James Reilly makes the first spacewalk from the Quest airlock in 2001. The airlock is depressurised and his feet stick out through the open hatch. *(NASA)*

space, brimming with stars. It is vast and overwhelming, this visual immersion, and I could drink it in forever."

As the spacewalkers carefully set off along hand rails to conduct their tasks, moving their personal tethers as they go, they remain conscious of the spacewalker's motto: "Always *make* a connection before you *break* a connection."

BELOW Astronaut in the Quest Crew Lock after dressing in the Equipment Lock. The Node Hatch leads to the main part of the ISS. *(NASA/David Woods)*

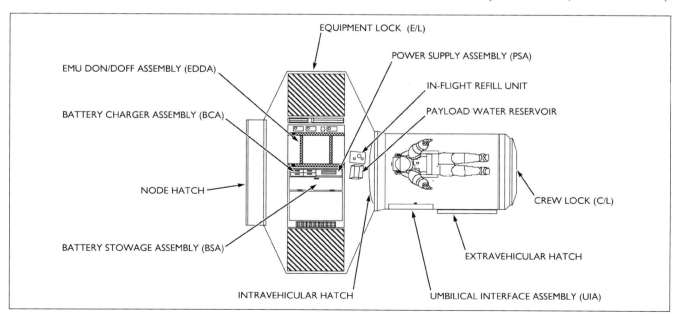

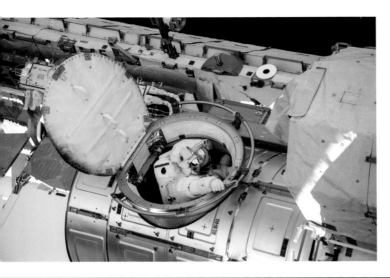

ABOVE LEFT NASA astronaut Chris Cassidy reaches out of the Quest airlock to clip his personal tether on to the safety bar. The flexible outer thermal hatch cover is stained with exposure to the space environment.

ABOVE ISS Expedition 35 spacewalkers Chris Cassidy (right) and Tom Marshburn replace a leaking pump controller box on the station's truss in 2013. They are wearing SAFER units 3 and 5 to permit a return to the station should they drift away.

LEFT Robert Stewart in the free-flying Manned Manoeuvring Unit off Space Shuttle *Challenger* on STS-41B.

OPPOSITE TOP During the STS-116 mission, Robert Curbeam (with red stripes) and Christer Fuglesang work above Cook Strait, New Zealand.

OPPOSITE BOTTOM ISS Expedition 37 Flight Engineer Sergey Ryazansky during a spacewalk in the Russian Orlan suit, in 2013. *(NASA)*

Chapter Nine

Back to Earth

◖─●─────────────◗

After a mission of some weeks or months, the astronaut returns to Earth in a fiery descent inside a small space capsule. Long days of carefree weightlessness come to an abrupt end in a crushing deceleration that peaks with five times the force of normal Earth gravity, then there is a violent jerk as the parachute opens, a pendulous descent, and the sharp kick of retro-rockets at touchdown. Ground crews open the hatch and the fragrant air of Earth is finally smelled once more.

OPPOSITE Only minutes from touchdown, the Descent Module of the Soyuz TMA-14M spacecraft hangs suspended from its single parachute above the clouds of Kazakhstan at dawn. It carries Soyuz commander Alexander Samokutyayev, Expedition 42 commander Barry Wilmore, and cosmonaut flight engineer Elena Serova. *(NASA/Bill Ingalls)*

159

Since the dawn of human spaceflight in 1961 there have been more than 300 crewed flights to Earth orbit, and at the end of each mission the crew have strapped themselves into their capsule and prepared themselves, both technically and mentally, for the most perilous part of the voyage.

The statistics of space history show that, while one crew was lost during launch, three crews never made it back to Earth alive. Eleven astronauts and cosmonauts died in those re-entry accidents, as compared to seven who perished on the way up to orbit.

Losing speed

A spacecraft orbiting the Earth has to shed a little over 16,000mph (26,000kph) in order to make a safe landing on the home planet. That arises from its orbital velocity of approximately 17,150mph (27,600kph), minus the speed of rotation of the Earth at the latitude of the landing point. That is 1,040mph (1,674kph) at the equator, and steadily declines as the latitude increases towards either pole.

Apollo 16 splashed down in the Pacific Ocean less than 0.5° south of the equator, and other American spacecraft splashed down to varying extents north and south. Russian spacecraft usually land in Kazakhstan at about 50°N, and the Chinese Shenzhou capsule lands in Inner Mongolia at around 40°N.

Retro-rockets can provide only a small proportion of the braking needed to lose that much speed. They fire in orbit against the direction of travel to depress the lowest point of the orbit, or perigee, sufficiently so that half an orbit later the spacecraft will encounter the upper layers of the atmosphere, and friction will do the rest. On Earth, this happens at about 75 miles (120km) altitude, which is above the Kármán Line. For future landings on Mars, with its less substantial atmosphere, this will occur at a height of 50 miles (80km) above the planet.

At the point known as the *entry interface*, the spacecraft is sufficiently braked that it cannot continue on its orbit, and can only plunge deeper into the atmosphere, accompanied by mounting frictional heating.

Gaining heat

The main challenge for spacecraft designers has been to protect the craft and its occupants from the heat of atmospheric re-entry, when temperatures exceeding 1,650°C (3,000°F) can be experienced.

Traditional capsules have used an ablative coating of a heat resistant material which slowly burns off, and as long as it is sufficiently thick at the points of maximum heating, it will not burn through to expose the structure beneath.

The US Space Shuttle used a complicated system of non-ablative tiles which protected the spaceplane from frictional heating, but did not burn off, so the craft was entirely reusable. Nevertheless, it did occasionally lose tiles and suffer heat damage to the metal airframe underneath. In the worst case, damage caused to a large protective tile on the leading edge of one wing during the ascent to orbit resulted in the destruction of *Columbia* during re-entry and the death of its crew over US territory in 2003.

Several capsule designs are currently being tested for future space operations, and all have a non-ablative heat protection system for re-entry in order to be reusable.

Deceleration loads

Once re-entry commences, the ionisation of hot gases around the spacecraft causes a blackout of radio communications, and the occupants are on their own for a few minutes. The deceleration is rapid, and the crew are pressed back into their seats at several times the force of gravity. This load can be alleviated if the shape of the capsule provides sufficient aerodynamic lift to steer a gentler trajectory.

The early spherical Russian capsules flew a purely ballistic trajectory which subjected the occupants to 7g. Occasionally, Soyuz and Shenzhou capsules may default to an unplanned ballistic descent that comes down well short of the targeted landing site and gives the crew a much rougher ride home.

The US Space Shuttle was retired following the landing of STS-135 *Atlantis* in 2011, so until the advent of the new American crew capsules that are currently being tested, the only way

home from the ISS is the venerable and reliable Russian Soyuz spacecraft.

Preparations for Soyuz return

If the crew have been in space for many months, some refresher training is advisable. This is done remotely with the Mission Control Centre near Moscow using an on-board simulator. They receive briefings on the precise timelines calculated for the specific orbit from which they will descend, and on conditions at the landing site. The crew rehearse the key stages of the return flight and the scenarios for an emergency descent, should this be required.

The Soyuz systems are activated and tested, as they may have been dormant for some weeks. Helen Sharman, the first British cosmonaut in

ABOVE The ionisation trail of the re-entry of Space Shuttle *Atlantis* on its final mission in 2011 is seen from the ISS. The Earth is illuminated by moonlight and the coloured band above the horizon is airglow, about 50 miles (80km) high.

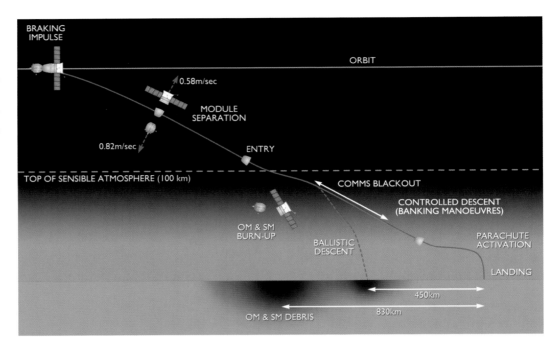

RIGHT Descent
path of the Soyuz
spacecraft from
orbit. The steeper
and rougher ballistic
descent route is only
followed in the event
of an anomaly.
(David Woods)

BRAKING
IMPULSE

ORBIT

0.58m/sec

MODULE
SEPARATION

0.82m/sec

ENTRY

TOP OF SENSIBLE ATMOSPHERE (100 km)

COMMS BLACKOUT

CONTROLLED DESCENT
(BANKING MANOEUVRES)

OM & SM
BURN-UP

BALLISTIC
DESCENT

PARACHUTE
ACTIVATION

LANDING

450km

830km

OM & SM DEBRIS

1991, recalls doing breathing exercises and a
pressure check on her lungs in preparation for
the forces she would experience on her return to
Earth. Tim Peake, the second UK space traveller
in 2016, also flew on a Russian craft and as he
prepared to depart for Earth on Soyuz TMA-19M
after 186 days of weightlessness he tweeted:
"Time to put on some weight!"

The departing crew say their goodbyes to
the cosmonauts who are remaining on the
ISS and then board their vehicle. Once the
Soyuz hatch with attached docking probe is in
place, the station crew close the ISS hatch that
receives the Soyuz probe. Seals are checked,
and the cosmonauts dress in their Sokol
pressure suits inside the Orbital Module before
transferring one by one to the Descent Module.

Soyuz undocking

The eight clamps that bind the docking ring
of Soyuz on to the ISS docking port are
opened, then devices are released to gently
push the Soyuz away at a rate of about 6 inches
(15cm) per second. The thrusters on the Soyuz
are not used close to the station, lest they
contaminate its exterior with chemical residue.
The craft slowly drifts away while the course is
monitored with external television cameras.

When it has retreated to about 70ft (20m),
a 15-second burst of thrusters increases

the separation rate to about 1.3mph (2kph).
Starting from the 250 miles (400km) altitude of
the ISS, the journey back to Earth will take a
little over three hours from undocking.

De-orbit engine firing

Soyuz is turned so that the main engine in
the rear of the Service Module is pointed
in the direction of travel, and a cover is opened
to expose the nozzle. In firing the rocket against
the craft's motion to bring it home on a nominal
trajectory, a specific change in velocity, referred
to as a Delta-V (ΔV), of 128m/s (285mph,
460kph) is required. This means burning the
engine for a duration of about 4min 45sec.

The effect of this manoeuvre is to lower
the perigee, or lowest point of the orbit on
the opposite side of the Earth, down into the
atmosphere. If there is too little deceleration, the
capsule will pass through the tenuous upper
atmosphere, slow a little, and proceed back out
into space on an unplanned orbit; too much
deceleration will steepen the re-entry path, and
subject the crew to additional *g*-forces and
dangerous overheating.

As the de-orbit engine burn proceeds, the
Soyuz commander calls out readings of Delta-V
achieved and burn time elapsed, such as
96m/s at 3min 30sec, 110m/s at 3min 59sec,
and so on. In the case of Soyuz TMA-19M, the

BELOW The remote
Kazakh town of
Arkalyk has been
the location of many
Russian spacecraft
landings. Its coat
of arms shows a
capsule descending
by parachute towards
ears of wheat, flanked
by ploughed furrows.

total retro manoeuvre of 149m/s was completed on time by a burn lasting 4min 37sec. On a nominal mission, the landing will occur about 50 minutes after the de-orbit burn.

Soyuz descends gradually, taking half an orbit or about 40 minutes to reach the entry interface. It is still in space, but on a track which will intercept the upper layers of the atmosphere, where air drag will ensure further deceleration and an inevitable descent to Earth, without any need for additional engine firings.

Spacecraft disassembly

As Soyuz drops steadily towards the atmosphere, several critical operations must be accomplished during its last minutes in space. The craft is carefully rotated so that the three modular compartments are aligned perpendicular to the direction of travel.

At an altitude of about 87 miles (140km), 20 minutes after the de-orbit burn and about 30 minutes before landing, explosive bolts almost simultaneously blow off the Orbital and the Service Modules, representing two-thirds of the weight of Soyuz. The periscope used to provide a forward view from the Descent Module is also discarded. Italian astronaut Paolo Nespoli returned from the ISS on Soyuz TMA-20 in 2011, and he recalls that the crew experience this separation phase as if a flurry of sledgehammer blows were raining down on different parts of the capsule from outside. The jettisoned components drift away on separate tracks that preclude their colliding with the Descent Module.

The Orbital Module, with its extra living space and rendezvous and docking equipment, is no longer required, and the Service Module with its solar arrays and main engine is not designed to return to Earth. Without the Service Module, the main navigation system and orientation jets are gone.

ABOVE A typical Soyuz re-entry track passes over southern Russia to land near the small town of Arkalyk in Kazakhstan, north of the Baikonur launch site. The cosmonauts are then transferred to Karaganda in preparation for their trip home. *(David Woods/Map from National Geographic's MapMaker interactive)*

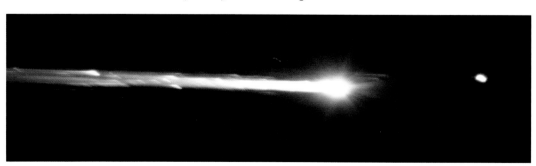

LEFT The Service and Orbital Modules of Soyuz TMA-05M burn up on re-entry over Kazakhstan in 2012, as seen from the ISS. *(NASA)*

The Descent Module, with the crew inside, is now on its own, but it contains secondary guidance, navigation and control systems which enable the crew to continue to manoeuvre the vehicle. The Soyuz commander pilots the capsule using a rotational hand controller that manages the firing of eight hydrogen peroxide thrusters. He orientates it so that the heat shield faces the direction of travel, to take the brunt of the re-entry. The crew continually consult their flight manuals to check operational procedures.

Re-entry

As the capsule penetrates the upper atmosphere and decelerates, the crew begin to experience a little g-force tugging at them, and weightlessness is finally over. This helps them to settle into their individually moulded seats and tighten their straps for the rough ride ahead. Floating objects drift to the rear of the cabin. This marks the entry interface, at an altitude of approximately 75 miles (120km).

Now, friction due to the thickening atmosphere begins to heat the outer surfaces of the capsule. Flames flicker at the windows, perhaps burning off the outer protective layer and alarming a first-time crew member who may be sitting next to it.

With only 20 minutes remaining before landing, the crew's attention turns to monitoring their descent path. The capsule has an offset centre of mass which creates aerodynamic lift. Control of the flight path can be achieved by rotating the craft and thereby changing the direction of this lift. This allows it to steer left or right and long or short to aim for the targeted landing site on the grassy steppes of Kazakhstan.

The discarded modules are also descending through the atmosphere on broadly similar but diverging tracks, but lacking protective heat shields they burn up in the atmosphere safely away from the crew capsule.

As the descent proceeds, the deceleration load on the crew increases to 1g, and then higher as the flames of re-entry flicker more brightly around the portholes. The normal maximum of 4g or 5g is felt at an altitude of about 22 miles (35km). The thick ablative material on the heat shield is now burning off as designed, and the blunt-end-first orientation generates a shock wave that stands off from the skin of the vehicle and thereby protects it from the most extreme temperatures.

The mounting g-forces are especially hard on cosmonauts who have been in space and weightless for several weeks or months. The deceleration continues to press them forcefully into their seats as they lie on their backs, with their knees up. The cosmonauts rehearse the rhythmic heavy breathing they have been taught, to keep their lungs inflated against the crushing weight of their chests. Time seems to pass slowly as they labour with these exercises, but finally their rate of descent falls below the sound barrier.

Parachutes

At an altitude of 34,500ft (6.5 miles or 10.5km), when the speed has dropped to 500mph (800kph), the noise of the wind rushing past is clearly heard.

The parachute cover is jettisoned with a bang, then the parachutes deploy. This process starts with two pilot chutes, the second one pulling out a drogue chute, which streams out on its cables and unfurls.

The drogue has an area of 258ft^2 (24m^2) and slows the rate of descent from 755ft (230m) to 262ft (80m) per second. This chute

BELOW Soyuz TMA-14M returns to Earth near the town of Zhezkazgan (Jezkazgan) in Kazakhstan carrying Alexander Samokutyaev, Elena Serova and Barry Wilmore. *(NASA)*

has a relatively small canopy diameter so it is not ripped apart by the high-speed airstream. Yet it still produces a very violent jerk, and the 2-tonne capsule tumbles in all directions, badly shaking the crew, until it settles down.

At an altitude of 28,000ft (5.3 miles, 8.5km), the drogue chute deploys the main parachute, which has an area of 10,800ft^2 (1,000m^2). Its harnesses shift the capsule's attitude to a 30° angle relative to the ground in order to dissipate the heat of re-entry from the shield. The single main parachute, ringed in red and white bands, finally slows the rate of descent to 15mph (24kph).

Inside the cabin, the crew know that most of the drama of the descent is now past, as they float downwards for a further ten minutes or so. The bangs and bumps are over and all that remains is the impact with the ground. Relieved to be safely through the worst, they will often smile and, despite the heaviness of their limbs, reach out and shake hands at this point. Paolo Nespoli recalls how heavy his head felt, and his wristwatch "weighed a tonne".

Meanwhile, the ground recovery crews, which are already deployed, are waiting and straining to spot the capsule in the sky.

The cosmonauts lie on their backs in normal Earth gravity, with the blue sky out of their portholes a welcome change from the blackness of space and the fire of re-entry.

The Soyuz has a single large descent parachute, unlike many other capsules such as Apollo, Crew Dragon and Blue Origin, which have three. Those three-chute capsules can survive the loss of one chute and still achieve a safe landing, but Soyuz has a back-up chute in case of problems with the main one. The back-up parachute is only 60 per cent the size of the main, with an area of 6,180ft^2 (574m^2), leading to a corresponding faster descent and greater impact upon landing.

Next, the heat shield on the base of the capsule and external window glass are jettisoned at an altitude of about 18,000ft (5,500m). The heat shield is still burning hot and accelerates downwards in a fast fall to Earth. The now-exposed shiny metal inner base of the capsule has retro-rockets on it. Out pops a 6ft (2m) metal probe which protrudes downward from the base of the capsule, with a ground sensor on the end, since replaced with the Kaktus altimeter.

Then the small attitude control jets that turned the craft in space for re-entry are de-activated, and their remaining fuel is jettisoned. This appears to outside observers like a stream of smoke trailing upwards from the descending craft.

The parachute harness adjusts again, this time to symmetric suspension to make a straight vertical descent for landing. In this

ABOVE **Soyuz TMA-14M descends into fog before landing in Kazakhstan with the crew of Expedition 42 after they spent six months at the ISS in 2014–15.** (NASA)

LEFT **Landing rockets cushion the impact of Soyuz TMA-08M carrying Expedition 36 commander Pavel Vinogradov, Alexander Misurkin and Chris Cassidy of NASA as they return to Earth in 2013.** (NASA)

RIGHT The Soyuz TMA-04M capsule disappears in a cloud of dust raised by its landing rockets near Arkalyk, Kazakhstan. It brought ISS Expedition 32 crew members Gennady Padalka, Joe Acaba and Sergei Revin home in 2012. *(NASA)*

BELOW Base of the Soyuz TM-7 capsule on display in the Memorial Museum of Cosmonautics, Moscow. In case it lands off course, there are instructions in Russian and English on how to open the crew hatch. It brought Alexander Volkov, Sergei Krikalyov and Valeri Polyakov back from the Mir space station in 1989. *(K. MacTaggart)*

configuration the cosmonauts' seats are lying horizontally to absorb the shock of impact.

The capsule's flight at this stage is uncontrolled in the horizontal direction, since it is carried on the wind and swings from side to side under the canopy, if the weather is gusty. Flight controllers have predicted for the ground recovery crew the approximate landing point based on what they know of the re-entry track and wind conditions, but it is not an exact science. Finally, someone scanning the skies with binoculars shouts that they have located the billowing chute. The ground and air observers now estimate the precise touchdown point. Helicopters are soon circling the descending capsule and directing ground vehicles towards the anticipated landing spot.

The cosmonauts feel a jerk as their

individually profiled seats are automatically elevated on struts to a brace position for the landing impact. They prepare for touchdown by folding their arms on their chest, keeping their tongue out from between their teeth, and nestling tightly down into the contoured body-form seat liners.

As the capsule floats down the final few metres towards the steppes, the extended probe touches the ground and triggers the final events. Six retro-rockets in the exposed base of the capsule fire downwards just 2ft (70cm) above ground level in a burst of flame, to slow the rate of descent to about 3mph (5kph). Inside the capsule, the cosmonauts experience a moment of violent drama. Despite the cushioning rocket firings, the capsule strikes the ground with a terrific thump and its occupants are thrown from side to side. The crushable seat supports absorb some of the force, and then there is silence. It is a moment of supreme relief for the crew, and the tension of the descent evaporates.

The approaching helicopter crews see a bright flash from the retro-rockets and then the capsule is briefly obscured by a thick cloud of smoke and dust. Released of its burden, the collapsing parachute flattens out into a pancake nearby.

Landed

After landing, the capsule can end up in two orientations – if there is little wind and the ground is flat, it will remain upright, with the access hatch on top. However, if a strong gust of wind catches the parachute before it collapses, it can tug the capsule over on to its side, and even drag it along the ground. In an effort to prevent the parachute tipping the capsule over, the Soyuz commander must now urgently release the chute harness.

As the Soyuz TMA-5 capsule rolled over, American astronaut Leroy Chiao found himself hanging in his seat directly above his crewmates in an awkward position that constricted an air tube in his suit. He was soon gasping for air, re-adapting to gravity and struggling to raise his colossally heavy arm to open his jammed helmet visor. About 80 per cent of Soyuz capsules end up on their sides.

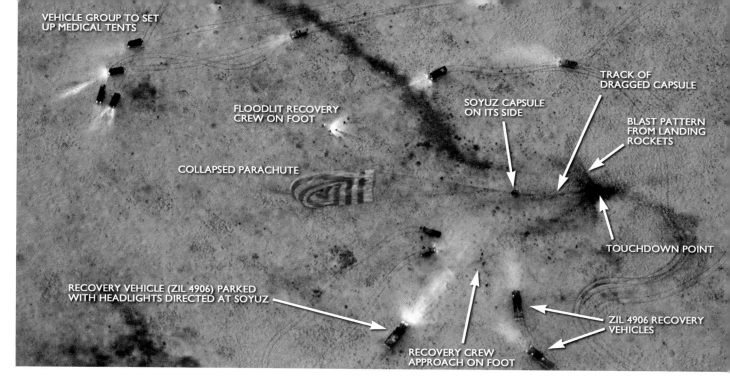

VEHICLE GROUP TO SET
UP MEDICAL TENTS

FLOODLIT RECOVERY
CREW ON FOOT

COLLAPSED PARACHUTE

RECOVERY VEHICLE (ZIL 4906) PARKED
WITH HEADLIGHTS DIRECTED AT SOYUZ

SOYUZ CAPSULE
ON ITS SIDE

TRACK OF
DRAGGED CAPSULE

BLAST PATTERN
FROM LANDING
ROCKETS

TOUCHDOWN POINT

ZIL 4906 RECOVERY
VEHICLES

RECOVERY CREW
APPROACH ON FOOT

The recovery troops land nearby and approach the still-hot capsule on foot. If it lies on its side, they might opt to manually roll it to the most comfortable angle, from which the crew can be most easily extracted. They look through the portholes and get their first glimpse of their charges. Ground crew attach a special tool to a screw in the centre of the hatch and rotate it open, and the crew finally breathe the fresh air of Earth.

Because an emergency landing may take place far away from the recovery team, or even abroad, the base of Soyuz, exposed by the jettisoned heat shield, bears stencilled instructions in Russian and English about how to extract the crew. A hatch-opening key is also provided in case the crew are incapable of opening it from the inside.

In the somewhat easier scenario of an upright landing, the capsule sits on its base with the crew hatch on top, and the cosmonauts have to be extracted upwards through the hatch and then lowered to the ground. To achieve this, the recovery crews have a metal frame with built-in ladders that straddles the top of the capsule and has legs extending to ground level on the sides. It creates a flat platform around the hatch for the

recovery personnel to work on, as they lift the cosmonauts one by one out into the open air. They sit there for a moment, getting their first whiff of the breezes and smells of Earth, and are then gently slid down a smooth metal chute on to the ground.

The ground crew arrange the trio side-by-side on reclining seats in the open air, where they have their initial medical checks. Helen Sharman also recalls a quick technical debriefing to ensure there had been nothing anomalous with the spacecraft's descent. After

ABOVE Soyuz
TMA-02M landed at
dawn near Arkalyk,
Kazakhstan in
November 2011. Its
crew of Sergey Volkov,
Michael Fossum and
Satoshi Furukawa
comprised Expedition
28 to the ISS.
(NASA/David Woods)

RIGHT Soyuz TMA-02M lies on its side at the
landing site in the aerial view shown above.
(NASA)

RIGHT Soyuz TMA-10M after landing with Expedition 38 commander Oleg Kotov, Mike Hopkins and Sergey Ryazansky in 2014. The capsule is straddled by steps leading to a platform via which crew members are extracted and slid down the chute on the right. *(NASA/Bill Ingalls)*

RIGHT Cosmonaut Sergey Ryazansky is extracted from Soyuz TMA-10M. *(NASA/Bill Ingalls)*

BELOW After their landing in Soyuz TMA-19M, the crew (from left) of Tim Peake, Yuri Malenchenko and Tim Kopra get a preliminary medical examination in the open air. *(ESA–Stephane Corvaja)*

a while they are transferred to a tent nearby for a fuller examination. Then there is the tradition of signing their names with chalk on the charred surface of their capsule.

Transfer from the landing zone

After the preliminaries, a cross-country vehicle drives the cosmonauts to the Mi-26 helicopters and each is flown in a separate helicopter to Karagandy (Karaganda) airport, a trip of about two hours. This means the risk of a helicopter delay or mishap will affect only one member of the Soyuz crew.

The helicopters are met by bus, which takes the three cosmonauts, now wearing more comfortable coveralls, to the terminal building.

BELOW Russian cosmonaut Elena Serova is well wrapped up after TMA-14M landed on a frosty March morning in 2015. *(NASA/Bill Ingalls)*

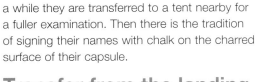

With a medical attendant supporting their arms on each side, they walk into the building and are greeted by three young women in colourful Kazakh national dress, who present each with a bunch of red roses. They receive the traditional Kazakh and Russian welcome of bread and salt.

Next they cross the hall to the medical centre for yet more check-ups, lasting up to an hour. When the all-clear is given, they emerge to sit in some smart chairs on top of a Kazakh carpet for a press conference. A table is decorated with the flags of Russia, Kazakhstan and the nations of the other astronauts flying on the mission. Some cosmonauts, especially first-timers, may be excused this ritual appearance if they are feeling queasy or out of balance.

Finally, a bus returns the cosmonauts to the apron, where two aircraft are waiting. One takes the Russian cosmonauts to Moscow, and a small NASA passenger jet is ready for American, Canadian and European astronauts. The NASA jet flies to Bodø in northern Norway to refuel, and here any European astronauts disembark and immediately board an ESA aircraft to fly to the ESA Astronaut Centre in Cologne, Germany.

Finally home with their families, the weary space travellers are subjected to further medical checks to enable researchers to collect more data on how their bodies and minds have adapted to living in space. At first they find beds uncomfortably hard, clothes are a drag on their movements, and they will occasionally leave a cup or pen in mid-air, only to be startled when it crashes to the floor! They will be monitored as they slowly readjust to gravity and deal with other aspects of normal terrestrial life.

RIGHT This Zil 4906 *Bluebird* amphibious vehicle carries a crane to recover the 3-tonne Soyuz capsule.

Chapter Ten

Life after space

Astronauts, like professional sportspeople, must usually find another occupation after their chosen career has peaked. For the most famous astronauts who are national heroes, life may pose some difficult challenges. Other astronauts have pursued a wide variety of post-flight work in aviation, engineering, business, public speaking, education and art.

OPPOSITE Alan Bean walked on the Moon in 1969 and later spent two months aboard the Skylab space station. He then left NASA to pursue a new career as an artist. He prides himself on being the only artist who paints scenes of the lunar surface from the perspective of one who has actually been there. His paintings capture significant moments during three years of human activity on the Moon, and express the wonder and joy of participating in an unparalleled act of exploration.

What to do in life after bearing the job title astronaut or cosmonaut can be a challenge, especially if the individual has accomplished something pioneering or unusual in space.

After becoming the first human in orbit, Yuri Gagarin toured the world and was rapturously received everywhere he went, even in the West despite the Cold War atmosphere of the time. His warm, lively personality made an enormous impression and attracted huge crowds on his early visits to London, Manchester, Czechoslovakia, Bulgaria, Finland, Italy and Germany. Later his trips to India, Sri Lanka, Japan, Egypt, Greece and other countries left a trail of monuments and commemorations which can still be seen. Of course, it would have been too much to visit the USSR's space rival, the USA, but he did travel to neighbouring Canada, Cuba and Brazil.

The restrictions of Soviet censorship were largely swept aside as Gagarin gave live interviews on BBC television and to the press, not as if he were boasting about Russia's space achievements to an ideological enemy, but simply explaining his experiences and his unique vision of the Earth. He met royalty, presidents and prime ministers, and had lunch with Queen Elizabeth II.

The crew of the first Moon landing, Neil Armstrong, Edwin 'Buzz' Aldrin and Michael Collins had an unusual return to Earth. As the first humans to visit another celestial body, there were real fears about the potential dangers of alien biological contamination, so they were sealed in strict quarantine for three weeks, being greeted by President Richard Nixon and their families through a small window.

Once the isolation was over, they also went on a world tour which lasted six weeks. Unlike Gagarin, they took their wives and covered 27 cities in 24 countries. Dazed by the rapidly changing venues, Collins quickly learned that it was safest to say "So nice to be in your lovely city", rather than risk offending their hosts by getting the location wrong.

None of those famous pioneers ever flew in space again, and their celebrity affected their lives quite dramatically. Owing to colossal media coverage, they were recognised everywhere they went. They were pestered by people wanting to shake their hands, demand their autographs and endlessly ask the same inane questions along the lines of: "What's it like up in space?" and "What was it like on the Moon?"

Armstrong was surprised to discover that his personal cheques were often not being cashed, because the recipient preferred to keep the signature of the first man on the Moon, rather than bank the money.

Gagarin was not allowed to return to space, at least initially, to his deep disappointment. At first he was grounded even from certain air flights, and had to travel by bus rather than car to reduce the likelihood that the Soviet Union's greatest modern hero would suffer a high speed accident. Years later, after lobbying his superiors, he was permitted to train for the early Soyuz program. The day he died in an air crash in 1968, he was flying in a MiG-15 trainer with instructor Vladimir Seryogin, clearly not because the ace space pilot needed any instruction in a basic air force jet, but because the Soviet authorities refused to risk him flying solo.

Coping with fame

Neither the world's first cosmonaut nor the first Moon landing crew were trained to cope with their sudden fame. As test pilots, military men and studious engineers, they were not accustomed to handling the various protocols involved in meeting kings, queens, shahs, presidents, popes, prime ministers and ambassadors. They were greeted by huge, intimidating crowds and had the overpowering attention of television hosts, politicians, actresses, newspapers and the general public lavished on them. Everyone, even the world's celebrities, wanted to meet the world's most celebrated spacemen.

Gagarin's wife soon complained that she saw more of her husband on television than she did in real life. In photos taken a few years after his flight, Gagarin does not look his best, sometimes appearing overweight and with puffy eyes. It must have been difficult to cope with soaring to global fame, and then being denied his chosen career of flying, instead being transformed into the perpetual role of a national and international hero, even a living legend.

The three Apollo 11 astronauts had varying experiences after their momentous flight to the Moon. Collins was offered the post of back-up commander of Apollo 14, which would have normally led on to the command of a landing three flights later, Apollo 17. That could have made him the last man on the Moon instead of Gene Cernan, but he declined to spend yet more years in tedious training. He left NASA in 1970 to become director of the National Air and Space Museum in Washington DC. Later, he quietly slipped into contented obscurity, writing and painting in Florida.

Neil Armstrong could have profited enormously from his fame with product endorsements, media appearances, public lectures or entering politics. Instead, having a retiringly modest but certainly not reclusive personality, he returned to academia as a professor of aerospace engineering at the University of Cincinnati, a quiet life that seemed to suit his temperament. He also built up various successful business interests, then retired to a farm in Ohio. He died during a medical operation in 2012.

ABOVE The Apollo 11 astronauts are greeted by a sea of humanity in Mexico City on their post-flight world tour. Neil Armstrong faces right, Michael Collins faces the camera, and a bewildered-looking Buzz Aldrin peers between them. *(NASA)*

The post-astronaut experiences of Edwin 'Buzz' Aldrin were in considerable contrast to his two crewmates. Buzz was severely shaken by the incessant attention and scrutiny of the world tour, and he openly admits that bouts of heavy drinking, depression and hospitalisation followed in the ensuing years. By 1977, desperate for a job, he was working as a used car salesman in Beverly Hills. He later came to see these

BELOW Valentina Tereshkova married fellow cosmonaut Andrian Nikolayev a few months after her spaceflight. *(RIA Novosti)*

RIGHT Despite international politics, the cosmonauts and astronauts got on well personally. Valentina Tereshkova and Georgy Beregovoy (left) welcome Neil Armstrong to Star City in 1970.

BELOW Valentina Tereshkova, here receiving an award from President Vladimir Putin, was elected to the Russian Duma.

setbacks as a critical point where he reassessed his life, and went on to change direction from his intense achievement-oriented career obsessions of an astronaut and Air Force officer to a more balanced outlook. He now uses his knowledge and experience to advocate for spaceflight, in particular for a mission to Mars.

A few months after she became the first woman in space, Valentina Tereshkova married fellow cosmonaut Andrian Nikolayev at a ceremony presided over by Soviet Premier Nikita Khrushchev. Their celebrity generated much stress, and they divorced after their daughter grew up. Later she hosted Neil Armstrong when he visited the cosmonaut training centre at Star City in 1970. Having shared the same dangers and experienced a unique perspective of Earth,

both nations' space flyers always greeted each other warmly. Visiting Gagarin's office, the first moonwalker wrote an elegant tribute to the first cosmonaut: "He called us all into space."

The randomness of history

The pressures of ultimate fame really only applied to those few astronauts who were chosen, often with a degree of randomness, for historic missions. Had a mishap befallen Gagarin in training or on the way to the launch pad, the iconic first man in space would have been Gherman Titov. Although rightly famous in space circles as the second man to orbit the Earth, and the first to spend a day in space, his recognition with the general public is negligible. Had Apollo 11 not quite worked out as planned, and the crew returned without landing, the first man on the Moon would have been the commander of Apollo 12, Pete Conrad, who had a jovial and outgoing personality.

Those astronauts whose names and faces are not etched in history generally experience relative anonymity after their space career is over. Bruce McCandless is inside the spacesuit in one of the most iconic images of the Space Age, a white-suited figure floating free with a jet backpack above the curve of the Earth, outlined against the blackness of space. Almost everyone in touch with the output of modern media will have seen the picture, which has

been endlessly reproduced, but the man himself can pass unrecognised almost anywhere.

The same is true of the pioneers of the Soviet space program – be it the first spacewalker Alexei Leonov, the first spacewoman Valentina Tereshkova or legendary spacewalker on Salyut space stations Georgi Grechko, who can all walk anonymously on a Moscow street. As the numbers of people who have flown in space inexorably rises, and their seemingly repetitive missions to the ISS become almost commonplace, some of the lustre is fading from the title 'astronaut'.

This will change for some future space achievements that we can anticipate today, such as the next nationality to follow the Americans to the Moon, and the first woman to walk there, and of course the first human on Mars. Historic stature awaits the first person born on Mars, but that is surely a century off.

National celebrities

The local exceptions to general anonymity tend to be the first, or in many cases the only, citizen of a nation to go into space – Helen Sharman of the UK, Yang Liwei of China, Dumitru-Dorin Prunariu of Romania, Sigmund Jähn of Germany (actually the former East German state, to the embarrassment of the western Federal Republic), Marc Garneau of Canada, the Icelandic-born Canadian Bjarni Tryggvason, Rodolfo Neri Vela of Mexico,

Sultan bin Salman Al Saud of Saudi Arabia, Rakesh Sharma of India, Abdul Ahad Momand of Afghanistan, and many more. However, even if their names may be familiar in their own countries, their faces are often much less so.

The late Sally Ride, who became the first American woman and the third female in space almost 20 years to the day after Valentina Tereshkova, is also widely remembered although she was something of a latecomer, being the 120th human to enter space.

A few astronauts never really left the profession. Many of the former Soviet cosmonauts continue to work and live at Star City outside Moscow, in training and management roles. Legendary American astronaut John Young who, over a period of 21 years commanded Gemini, Apollo and Space Shuttle missions and walked on the Moon, worked for NASA for 42 years, and even after retiring at the age of 74 he continued to attend NASA meetings and pursue other space projects.

Retirement occupations

Several space travellers have gone into politics. Valentina Tereshkova, the first woman in space, became a member of the Supreme Soviet after her flight, later serving in the Presidium until the dissolution of the USSR. Then in 2011 she was elected to the State Duma, the lower house of the Russian legislature. Vladimir Remek of Czechoslovakia,

whose nation was the third to have a citizen enter space after the USSR and USA, was elected to the European Parliament in 2004 and later became the Czech ambassador to Russia.

Geologist Harrison 'Jack' Schmitt flew on the Apollo 17 mission in 1972, becoming the first scientist on the Moon. He later became Senator for his home state of New Mexico. John Glenn, famous as the first American to orbit the Earth and, like Gagarin, a national treasure denied another spaceflight, was a Senator for Ohio for 24 years but failed in his presidential aspirations. After lobbying NASA and passing all medical tests, he returned to space in 1998 on the Shuttle *Discovery* as part of a study on human ageing, becoming by far the oldest person to go into orbit at the age of 77 years.

After his mission, the only Romanian cosmonaut, Dumitru-Dorin Prunariu, had several prominent jobs. He was appointed as his nation's ambassador to Russia, served as chairman of the UN Committee on the Peaceful Uses of Outer Space (COPUOS) and later worked for the European Space Agency.

After his Apollo 9 mission, Rusty Schweickart worked in various science and technology posts and co-founded the Association of Space Explorers (ASE). In 2002 he helped establish the B612 Foundation along

ABOVE Czech cosmonaut Vladimir Remek was elected to the European Parliament.

RIGHT Dr Harrison 'Jack' Schmitt, the first scientist on the Moon, became US Senator for New Mexico.

BELOW Almost 37 years after becoming the first American in orbit, John Glenn flew to space again and became the oldest astronaut at the age of 77.

BELOW Rusty Schweickart has campaigned to raise awareness of the danger posed to the Earth by asteroids. *(David Woods)*

ABOVE Schweickart helped establish the B612 Foundation, named after the fictional asteroid in *Le Petit Prince* by Antoine de Saint-Exupéry.

with former astronaut Ed Lu, also serving as its chair. This non-profit body is dedicated to studying the dangers of asteroid impacts and defending the Earth from them.

John 'Jack' Swigert was a back-up pilot during the preparations for Apollo 13, but found himself flying to the Moon at three days' notice after his counterpart Ken Mattingly was exposed to German measles. He was elected to the US House of Representatives in 1982, but died before he could take up his seat.

Other astronauts have risen to prominence in the business world. Having commanded Gemini VII in Earth orbit and Apollo 8 in lunar orbit, Frank Borman became Chief Executive Officer, then Chairman of Eastern Airlines based in Florida. Some Russian cosmonauts, including Alexei Leonov, served on the boards of banks and other businesses.

Many former astronauts capitalise on their experiences by turning to public speaking, consultancy and education. There is certainly no obvious or typical career route for them, and they will often assemble a collection of different activities into an occupation. American astronaut Leroy Chiao, for instance, has based his post-NASA career on what he learned in preparing for and flying three Space Shuttle flights and commanding Expedition 10 to the ISS from 2004 to 2005. He has been involved in business ventures in the US, acts as engineering consultant on space projects, holds university appointments, chairs a space medicine research institute panel and runs OneOrbit, a motivational, training and education business. Chiao has a special interest in the space ambitions of China, from where his family originates.

BELOW Frank Borman led the first mission to orbit the Moon, then became chief of Eastern Airlines.

Changed perspectives

Perhaps surprisingly for analytically minded pilots, engineers and scientists, several astronauts have turned to expressing themselves in art. Maybe the creative urge comes from the transforming vision that many of them had of their home planet, dissolving their preconceptions about human activities, nations and their borders, and the apparent permanence of our planetary environment. Instead, there emerges a deep appreciation of the Earth's fragility and insignificance in the cosmos.

Earth-orbiting astronauts perceive our habitable biosphere as the thinnest imaginable veneer draped around the globe, and surrounded by an inhospitable vacuum. Those few privileged individuals who made the voyage to the Moon saw a fragile blue and white jewel hanging in the lethal black void, so small that they could cover it with their thumb.

The day-long orbital mission of Yang Liwei in 2003 made China only the third nation to send its own astronaut into space. Yang later told Leroy Chiao, who was the first US astronaut to visit the Chinese space centre, that his biggest impression of space was the total absence of borders. This fresh and more authentic perspective seems to be a widespread impact of spaceflight on human consciousness.

Alexei Leonov started the artistic trend in space travellers, initially painting terrestrial scenes, then taking crayons into space on his Voskhod 2 mission in 1965 and making a colour sketch of the rising Sun while in orbit. He felt that it calmed him after his traumatic spacewalk. His pencils were attached by string to a rubber ring to stop them floating off in weightlessness. He also took his art materials on the joint Soyuz-Apollo mission with the Americans, making several drawings of the crew of both spacecraft as well as a self-portrait during the mission.

Back on Earth, he has produced many more artistic space scenes, some of them for use on postage stamps, and his works have been exhibited at several institutions including the National Air and Space Museum in Washington DC. He has published several books of paintings, including *The Sun's Wind*, and *Earth and Space Painting*.

Alan Bean took painting classes before he travelled to the Moon in 1969. After commanding the second Skylab space station

mission, he considered his future career. "I thought of continuing on to the Space Shuttle program, but I realised there were plenty of younger people who could do that. As an artist who had been to the Moon, I was in a unique position to record my own experiences and those of my fellow moon voyagers on Apollo." So he resolved to become a full-time artist, and has been so since the 1980s.

Bean first creates a sculpted plaster surface which he subtly marks with his Moon boot tread and geology tools, and embeds threads of his spacesuit patches which are dirty with real moon dust. Then he paints a lunar scene in acrylics, recording an actual incident from lunar exploration or an imaginative depiction of the spirit of Apollo. While there are many artists, some of them former astronauts, who render space scenes, Bean has a unique viewpoint conferred by the singular fact that he is the only artist to have walked on another world. He believes that one day a painting of his will hang in a gallery on the Moon, and he is surely right.

ABOVE *Moonrock – Earthbound.* **Moonwalker Alan Bean painted the Apollo 16 crew, John Young and Charlie Duke, collecting a rock at their Descartes landing site.** *(Alan Bean)*

BELOW **Astronaut Nicole Stott created this colourful artwork based on a spacesuit concept.**

More recently, Shuttle and ISS astronaut Nicole Stott flew in 2009 and 2011, spending 104 days in orbit and making a spacewalk on Shuttle *Discovery* flight STS-128. Stott relates that "on seeing the overwhelming beauty" of the Earth from space she "had an epiphany", and resolved to spend the rest of her life sharing the experience with others through her art, public speaking and teaching.

Ron Garan felt similarly after spending a total of 178 days in orbit on the Space Shuttle, Soyuz and the ISS between 2008 and 2011, and making several spacewalks. Garan was impressed by how 15 nationalities could collaborate to build and operate the ISS, one of the most ambitious and technologically complicated undertakings in history. In his post-astronaut life, he is applying this spirit of cooperation and innovation towards creating a better world, in what he calls the 'orbital perspective':

'I left my dream job as an astronaut, a career I worked my entire life to achieve. I left NASA for one overarching reason, to be able to share a unique perspective of our planet that I believe can have profound, positive effects on the trajectory of our global society and our world. I left NASA so I can share that perspective full time.'

Many astronauts wrote books about their extraordinary experiences, and some wrote several. Yuri Gagarin's *Road to the Stars* was published in many languages. Gemini and Apollo astronaut David Scott co-authored *The Other Side of the Moon* with legendary Soviet cosmonaut Alexei Leonov, to produce an extraordinary bridge between two Cold War rivals.

Walter Cunningham flew on the first successful Apollo flight, rising off the very launch pad where the crew of Apollo 1 had perished in a fire, and his book *The All-American Boys* is an honest, insider's account of being an astronaut in its heyday. Alfred Worden orbited the Moon and made the first deep space EVA in 1971, and he frankly recounts the story of his space career and its untimely termination in *Falling to Earth*. He also published a small book of his own space-inspired poetry entitled *Hello Earth*.

The first moonwalker Neil Armstrong never wrote his autobiography, but he did authorise James Hansen to compile the definitive account of his life in *First Man*. Published in 2005, it is

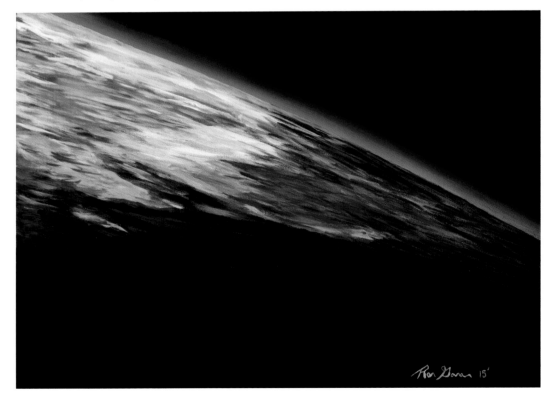

RIGHT Ron Garan flew on the Shuttle, Soyuz and ISS: "In this piece, *Crossing the Terminator*, I tried to capture the emotion that I felt as I watched the colours change on Earth in a blur of movement and speed." *(Ron Garan)*

being made into a biopic by Universal Pictures, with Ryan Gosling in the starring role.

Many of the books written in retirement by Russian cosmonauts are not so well known but are no less interesting, especially now that the former restrictions of Soviet censorship have fallen away, and the epic flights of the pioneers of space can be fully exposed. Alexei Eliseev, a three-flight veteran from the early Soyuz program, wrote *Life – A Drop in an Ocean* (2001). The frank *Diary of a Cosmonaut: 211 Days in Space* by Valentin Lebedev recounts his record-setting time on the space station Salyut 7.

Michael Collins reflected on life after space in *Carrying the Fire* (1974), which is one of the best autobiographical accounts ever written by an astronaut:

'Being an astronaut was the most interesting job I ever expect to have, but I wanted to leave before I became stale in it, and I could tell that after Apollo 11 I could not have prevented myself from sliding downhill, in terms of enthusiasm and concentration. Hence I find myself in the weird position of saying I'm glad I no longer have one of the most fascinating jobs in the world…'

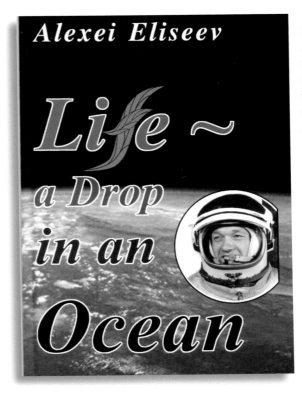

LEFT Cosmonaut Alexei Eliseev wrote a fascinating inside account of the early Soviet space program. *(Incombook)*

BELOW The Apollo 11 crew (from left) Buzz Aldrin, Michael Collins and Neil Armstrong at the White House to mark the 40th anniversary of their Moon flight, 20 July 2009.

Glossary

ACES	Advanced Crew Escape Suit (US).
Apollo	The US Moon-landing program, which also featured preliminary flights in Earth and Moon orbit, 1968-72.
ASTP	Apollo-Soyuz Test Project.
CCP	Commercial Crew Program (US), which promotes private sector access to space.
CNSA	China National Space Administration, the national space agency of the People's Republic of China.
CSA	Canadian Space Agency.
EMU	Extravehicular Mobility Unit, the modern US spacesuit plus its backpack.
ESA	European Space Agency.
EVA	Extra-Vehicular Activity, a spacewalk or Moon walk.
FAA	Federal Aviation Administration, the US body regulating commercial spaceflight.
g	The force of gravity as experienced on the Earth's surface.
Gemini	The second USA manned space program, for pairs of astronauts, flew 1965-66.
Gnomon	Pointer which casts a shadow. A photo-calibration gnomon with colour chart was used to document the collection of lunar samples on Apollo missions.
HUT	Hard Upper Torso, top part of the SSA (US).
ISS	International Space Station (see also MKS).
JAXA	Japan Aerospace Exploration Agency.
JSC	Johnson Space Center, Houston, USA.

Kármán Line	Internationally-agreed boundary of space, altitude 100km (62 miles).
Launch Vehicle	A rocket which propels a spacecraft into space.
LTA	Lower Torso Assembly (US), the lower half of the SSA.
MACES	Modified Advanced Crew Escape Suit (US) – see also ACES.
Mercury	The first USA manned space program, flew 1961-63.
Mir	Modular Soviet/Russian space station, 1986-2001.
MKS	Russian abbreviation for the International Space Station (Mezhdunarodnaya Kosmicheskaya Stantsiya).
MMU	Manned Maneuvering Unit, the jet backpack flown from the US Space Shuttle.
MSFC	Manned Space Flight Center, Houston, now called the JSC.
NACA	National Advisory Committee for Aeronautics (US), founded 1915 and dissolved into NASA in 1958.
NASA	National Aeronautics and Space Administration (US).
Orbit	Curved path, usually elliptical, of an object (e.g. rocket, satellite or spacecraft) around an astronomical body such as the Earth, Moon, Sun or a planet.
PLSS	Primary Life Support Subsystem (USA), worn as a backpack during EVAs.
Progress	Unmanned Russian supply-vessel for ISS and previous space stations, first used in 1978.
Roskosmos	Russian Federal Space Agency.

SAFER	Simplified Aid For EVA Rescue. A slim jet-pack fitted to the PLSS during an EVA, so an astronaut can fly back to the ISS if they become detached. It is only activated in an emergency.
Salyut	Series of USSR (Russian) space stations, crewed 1971-86.
Shenzhou	China's crewed spacecraft, which takes up to 3 people.
Skylab	The first US space station which hosted three crews, 1973-74.
Soyuz	Current Russian manned spacecraft type, first flown with crew by the USSR in 1967.
Soyuz FG	The launch vehicle (rocket) which puts the Soyuz spacecraft into orbit. The FG model first flew in 2001, and has launched Soyuz TMA and Soyuz MS craft.
Soyuz MS	Final Soyuz spacecraft variant. First flew in 2016.
Soyuz T	Soyuz Transport. Flown 1979-86.
Soyuz TM	Soyuz Transport Modified. Flown 1986-2002.
Soyuz TMA	Soyuz Transport Modified Anthropometric. Anthropometric refers to the fact that interior fittings are adaptable to the statures of individual crew members. Flown 2002-11.
Soyuz TMA-M	Further modification of Soyuz TMA type. Flown 2010-16.
SpaceShipOne	Experimental sub-orbital spaceplane, designed by Burt Rutan of Scaled Composites, which became the first non-government piloted craft to reach space.
SpaceShipTwo	Passenger-carrying sub-orbital spacecraft, developed from SS1, designed to take tourists on short trips into space.

Space Shuttle	US reusable crewed spacecraft type, flew 135 missions 1981-2011.
SS1	Abbreviation for SpaceShipOne.
SS2	Abbreviation for SpaceShipTwo.
SSA	Space Suit Assembly (US). When fitted with the PLSS backpack it becomes the EMU.
STS	Space Transportation System, full name for the US Space Shuttle.
Sub-orbital	A ballistic flight with insufficient velocity to achieve orbit, even if it enters space.
Tiangong	China's series of space stations. First one launched in 2011.
Vostok	The first manned Russian (USSR) spacecraft, flew 1961-63.
Voskhod	Multi-person Russian (USSR) spacecraft based on a modified Vostok. It only flew twice, first with a crew of three in 1964, then with a crew of two and an airlock for spacewalking, in 1965.
WhiteKnightOne	Carrier aircraft for SS1 sub-orbital spaceplane.
WhiteKnightTwo	Carrier aircraft for SS2 sub-orbital spaceplane.
WK1	Abbreviation for WhiteKnightOne.
WK2	Abbreviation for WhiteKnightTwo.
X-15	US rocket-powered aircraft which made sub-orbital spaceflights in 1963.
USSR	Union of Soviet Socialist Republics (or Soviet Union), predecessor state to modern Russia. It started the USSR/Russian space program with the launch of Sputnik in 1957. Dissolved in 1991.

Index

US Space Shuttle *Endeavour* heads for Earth orbit from NASA's Kennedy Space Center on STS-134, the penultimate mission of the program, in May 2011. It took the six-member crew of Mark Kelly, Greg Johnson, Michael Fincke, Andrew Feustel, Greg Chamitoff and Italian astronaut Roberto Vittori to a rendezvous with the International Space Station. The photograph was taken from a Shuttle training aircraft. *(NASA)*